International Vocational Education Bilingual Textbook Series
国际化职业教育双语系列教材

The Technology of Secondary Refining
炉外精炼技术

Zhang Zhichao
张志超　编

Beijing
Metallurgical Industry Press
2020

内 容 提 要

本书共分三个项目：项目1主要介绍炉外精炼手段与工艺选择，包含5个任务，内容主要涉及合成渣洗、搅拌、加热、真空和喷吹；项目2主要介绍炉外精炼基本工艺，包含13个任务，内容主要涉及LF工艺、LF精炼造渣技术、LF泡沫渣精炼工艺、典型钢种LF精炼工艺、LF炉精炼设备、ASEA-SKF钢包精炼、CAS精炼、VD精炼、VAD精炼与VOD精炼、DH精炼、RH精炼、RH精炼和AOD精炼；项目3主要介绍炉外精炼用耐火材料基础知识。

本书可作为职业院校冶金相关专业的国际化教学用书，也可作为冶金企业员工的培训教材和有关专业人员的参考书。

图书在版编目(CIP)数据

炉外精炼技术 = The Technology of Secondary Refining：汉、英 / 张志超编．—北京：冶金工业出版社，2020.8
国际化职业教育双语系列教材
ISBN 978-7-5024-8537-5

Ⅰ.①炉… Ⅱ.①张… Ⅲ.①炉外精炼—高等职业教育—双语教学—教材—汉、英 Ⅳ.①TF114

中国版本图书馆 CIP 数据核字(2020)第 152087 号

出版人 陈玉千
地　址　北京市东城区嵩祝院北巷39号　邮编　100009　电话　(010)64027926
网　址　www.cnmip.com.cn　电子信箱　yjcbs@cnmip.com.cn
责任编辑　俞跃春　刘林烨　美术编辑　郑小利　版式设计　孙跃红　禹 蕊
责任校对　郑　娟　责任印制　李玉山
ISBN 978-7-5024-8537-5
冶金工业出版社出版发行；各地新华书店经销；三河市双峰印刷装订有限公司印刷
2020年8月第1版，2020年8月第1次印刷
787mm×1092mm　1/16；14.5 印张；347 千字；210 页
56.00 元

冶金工业出版社　　投稿电话　(010)64027932　投稿信箱　tougao@cnmip.com.cn
冶金工业出版社营销中心　电话　(010)64044283　传真　(010)64027893
冶金工业出版社天猫旗舰店　yjgycbs.tmall.com

(本书如有印装质量问题，本社营销中心负责退换)

Editorial Board of International Vocational Education Bilingual Textbook Series

Director　　Kong Weijun (Party Secretary and Dean of Tianjin Polytechnic College)

Deputy Director　　Zhang Zhigang (Chairman of Tiantang Group, Sino-Uganda Mbale Industrial Park)

Committee Members　　Li Guiyun, Li Wenchao, Zhao Zhichao, Liu Jie, Zhang Xiufang, Tan Qibing, Liang Guoyong, Zhang Tao, Li Meihong, Lin Lei, Ge Huijie, Wang Zhixue, Wang Xiaoxia, Li Rui, Yu Wansong, Wang Lei, Gong Na, Li Xiujuan, Zhang Zhichao, Yue Gang, Xuan Jie, Liang Luan, Chen Hong, Jia Yanlu, Chen Baoling

国际化职业教育双语系列教材编委会

主　任　孔维军（天津工业职业学院党委书记、院长）

副主任　张志刚（中乌姆巴莱工业园天唐集团董事长）

委　员　李桂云　李文潮　赵志超　刘　洁　张秀芳
　　　　　谭起兵　梁国勇　张　涛　李梅红　林　磊
　　　　　葛慧杰　王治学　王晓霞　李　蕊　于万松
　　　　　王　磊　宫　娜　李秀娟　张志超　岳　刚
　　　　　玄　洁　梁　娈　陈　红　贾燕璐　陈宝玲

Foreword

With the proposal of the 'Belt and Road Initiative', the Ministry of Education of China issued *Promoting Education Action for Building the Belt and Road Initiative* in 2016, proposing cooperation in education, including 'cooperation in human resources training'. At the Forum on China-Africa Cooperation (FOCAC) in 2018, President Xi proposed to focus on the implementation of the 'Eight Actions', which put forward the plan to establish 10 Luban Workshops to provide skills training to African youth. Draw lessons from foreign advanced experience of vocational education mode, China's vocational education has continuously explored and formed the new mode of vocational education with Chinese characteristics. Tianjin, as a demonstration zone for reform and innovation of modern vocational education in China, has started the construction of 'Luban Workshop' along the 'Belt and Road Initiative', to export high-quality vocational education achievements.

The compilation of these series of textbooks is in response to the times and it's also the beginning of Tianjin Polytechnic College to explore the internationalization of higher vocational education. It's a new model of vocational education internationalization by Tianjin, response to the 'Belt and Road Initiative' and the 'Going Out' of Chinese enterprises. Tianjin Polytechnic College and Uganda Technical College-Elgon reached a cooperation intention to establish the Luban Workshop to carry out vocational education cooperation on mechatronics technology and ferrous metallurgy technology major in 2019. The establishment of Luban Workshop is conducive to strengthen the cooperation between China and Uganda in vocational education, promote the export of high-quality higher vocational education resources, and serve Chinese enterprises in Uganda and Ugandan local enterprises. Exploring and standardizing the overseas operation of Chinese colleges, the expansion of international influences of China's higher vocational education is also one of the purposes.

The construction of 'Luban Workshop' in Uganda is mainly based on the EPIP (Engineering, Practice, Innovation, Project) project, and is committed to cultivating high-quality talents with innovative spirit, creative ability and entrepreneurial spirit. To meet the learning needs of local teachers and students accurately, the compilation of these international vocational skills bilingual textbooks is based on the talent demand of Uganda and the specialty and characteristics of Tianjin Polytechnic.

These textbooks are supporting teaching material, referring to Chinese national professional standards and developing international professional teaching standards. The internationalization of the curriculums takes into account the technical skills and cognitive characteristics of local students, to promote students' communication and learning ability. At the same time, these textbooks focus on the enhancement of vocational ability, rely on professional standards, and integrate the teaching concept of equal emphasis on skills and quality. These textbooks also adopted project-based, modular, task-driven teaching model and followed the requirements of enterprise posts for employees.

In the process of writing the series of textbooks, Wang Xiaoxia, Li Rui, Wang Zhixue, Ge Huijie, Yu Wansong, Wang Lei, Li Xiujuan, Gong Na, Zhang Zhichao, Jia Yanlu, Chen Baoling and other chief teachers, professional teams, English teaching and research office have made great efforts, receiving strong support from leaders of Tianjin Polytechnic College. During the compilation, the series of textbooks referred to a large number of research findings of scholars in the field, and we would like to thank them for their contributions.

Finally, we sincerely hope that the series of textbooks can contribute to the internationalization of China's higher vocational education, especially to the development of higher vocational education in Africa.

Principal of Tianjin Polytechnic College Kong Weijun
May, 2020

序

随着"一带一路"倡议的提出，2016年中华人民共和国教育部发布了《推进共建"一带一路"教育行动》，提出了包括"开展人才培养培训合作"在内的教育合作。2018年习近平主席在中非合作论坛上提出，要重点实施"八大行动"，明确要求在非洲设立10个鲁班工坊，向非洲青年提供技能培训。中国职业教育在吸收和借鉴发达国家先进职教发展模式的基础上，不断探索和形成了中国特色职业教育办学模式。天津市作为中国现代职业教育改革创新示范区，开启了"鲁班工坊"建设工作，在"一带一路"沿线国家搭建"鲁班工坊"平台，致力于把优秀职业教育成果输出国门与世界分享。

本系列教材的编写，契合时代大背景，是天津工业职业学院探索高职教育国际化的开端。"鲁班工坊"是由天津率先探索和构建的一种职业教育国际化发展新模式，是响应国家"一带一路"倡议和中国企业"走出去"，创建职业教育国际合作交流的新窗口。2019年天津工业职业学院与乌干达埃尔贡技术学院达成合作意向，共同建立"鲁班工坊"，就机电一体化技术专业、黑色冶金技术专业开展职业教育合作。此举旨在加强中乌职业教育交流与合作，推动中国优质高等职业教育资源"走出去"，服务在乌中资企业和乌干达当地企业，探索和规范我国职业院校"鲁班工坊"建设和境外办学，扩大中国高等职业教育的国际影响力。

中乌"鲁班工坊"的建设主要以工程实践创新项目（EPIP：Engineering, Practice, Innovation, Project）为载体，致力于培养具有创新精神、创造能力和创业精神的"三创"复合型高素质技能人才。国际化职业教育双语系列教材的编写，立足于乌干达人才需求和天津工业职业学院专业特色，是为了更好满足当地师生学习需求。

本系列教材采用中英双语相结合的方式，主要参照中国专业标准，开发国际化专业教学标准，课程内容国际化是在专业课程设置上，结合本地学生的技术能力水平与认知特点，合理设置双语教学环节，加强学生的学习与交流能力。同时，教材以提升职业能力为核心，以职业标准为依托，体现技能与质量并重的教学理念，主要采用项目化、模块化、任务驱动的教学模式，并结合企业岗位对员工的要求来撰写。

本系列教材在撰写过程中，王晓霞、李蕊、王治学、葛慧杰、于万松、王磊、李秀娟、宫娜、张志超、贾燕璐、陈宝玲等主编老师、专业团队、英语教研室付出了辛勤劳动，并得到了学院各级领导的大力支持，同时本系列教材借鉴和参考了业界有关学者的研究成果，在此一并致谢！

最后，衷心希望本系列教材能为我国高等职业教育国际化，尤其是高等职业教育走进非洲、支援非洲高等职业教育发展尽绵薄之力。

天津工业职业学院书记、院长　孔维军
2020 年 5 月

Preface

Tianjin Polytechnic College and Uganda Technical College-Elgon reached a cooperation intention to establish the Luban workshop to carry out vocational education cooperation on mechatronics technology and ferrous metallurgy technology major in 2019. In order to strengthen the cooperation between China and Uganda in vocational education, the two colleges plan to compile a series of international vocational skills bilingual textbooks.

This book is one of the international vocational skills bilingual textbooks, mainly aiming at steel-making production. From the aspects of strengthening the quality of molten steel, optimizing the production organization and reducing the production cost, it introduces the commonly used secondary refining in the steel-making process in China and the world. From the principle of technology to the use of equipment, they are all compiled according to the production technology used by enterprises at present. In addition, combined with the quality index of advanced production enterprises, the related production process and the refining of special steel are introduced. However, there are many kinds of secondary refining, the actual operation is complex, the production conditions and personnel quality of each enterprise are different, so it is impossible to have all the production technologies.

The book is written by Zhang Zhichao of Tianjin Polytechnic College, referencing to the relevant materials and literature in the process of writing. Here express my gratitude to the authors concerned.

Due to the limited level of the editor, there is something wrong in the book. I hope readers to criticize and correct.

The editor
May, 2020

前 言

2019年天津工业职业学院与乌干达埃尔贡技术学院达成合作意向，共同建立"鲁班工坊"，就机电一体化技术专业、黑色冶金技术专业开展职业教育合作，双方计划编撰国际化职业教育双语系列教材。

本书是国际化职业教育双语系列教材之一。本书主要针对炼钢生产，从强化钢水质量、优化生产组织、降低生产成本等方面，介绍了目前中国及世界炼钢生产工序中常用的炉外精炼技术。从技术原理到设备使用，都是依据目前企业所用的生产技术编写。另外，在结合先进生产企业的优质指标中，介绍了相关生产流程，以及特色钢种的炉外精炼。但炉外精炼种类繁多，实际操作复杂，各企业生产条件、人员素质都不同，因此不可能将所有的生产技术都面面俱到。

本书由天津工业职业学院张志超编写，在编写过程中，参考了有关资料和文献，在此向有关作者表示感谢。

由于编者水平所限，书中不妥之处，希望读者批评指正。

编 者
2020 年 5 月

Contents

0 Introduction ·· 1

  0.1 The Reasons for Development of Secondary Refining ································ 1

  0.2 Development History of Secondary Refining ··· 3

    0.2.1 Vacuum Treatment ·· 4

    0.2.2 Non Vacuum Treatment ·· 5

    0.2.3 Development of China's Secondary Refining ··· 8

    0.2.4 Role and Characteristics of Refining in Modern Steel Production ····················· 10

    0.2.5 The Main Purpose and Task of Secondary Refining ·· 14

    0.2.6 Selection of Refining Equipment ·· 15

    0.2.7 Development Trend of Secondary Refining ·· 17

Project 1 Refining Means and Process selection of Secondary Refining ······ 19

  Task 1.1 Synthetic Slag Washing ··· 21

    1.1.1 Physical and Chemical Properties of Synthetic Slag ·· 22

    1.1.2 Slag Washing Process ·· 27

    1.1.3 Application Technology ·· 30

    1.1.4 Effect and Deficiency ·· 31

  Task 1.2 Mixing ··· 34

    1.2.1 Mixing Method ··· 35

    1.2.2 Mixing and Homogenization ··· 43

  Task 1.3 Heating ·· 43

    1.3.1 Heating Method ·· 44

    1.3.2 Selection of Heating Process ··· 51

  Task 1.4 Vacuum ·· 52

    1.4.1 Vacuum and Its Utilization in Metallurgical Technology ······································ 54

    1.4.2 Vacuum Degassing of Liquid Steel ·· 56

1.4.3　Vacuum Deoxidation of Molten Steel ················· 63

Task 1.5　Injection ················· 67

Project 2　Basic Process of Secondary Refining ················· 71

Task 2.1　LF Process ················· 74

2.1.1　LF Production Process ················· 76

2.1.2　The Main Process System ················· 81

Task 2.2　LF Refining Slag Technology ················· 88

2.2.1　Properties of Synthetic Slag ················· 90

2.2.2　Removal of Inclusions ················· 92

2.2.3　Desulfurization Process ················· 93

Task 2.3　LF Foam Slag Refining Process ················· 103

2.3.1　Foam Slag Metallurgical Effect ················· 104

2.3.2　Slag Foaming ················· 104

2.3.3　Foaming Agent for LF Submerged Arc Refining Slag ················· 109

2.3.4　Impact of Operation Process ················· 114

2.3.5　Detection and Control Technology of Foam Residue ················· 116

Task 2.4　Typical Steel LF Refining Process ················· 117

2.4.1　General Carbon and Low Alloy Refining Process ················· 117

2.4.2　Refining Process of Medium and High Carbon Steel ················· 118

2.4.3　Refining Process of High Quality Plate ················· 119

2.4.4　Slag for Refining Low Carbon Aluminum Containing Steel ················· 120

2.4.5　Smelting of Bearing Steel ················· 121

2.4.6　Other New Technologies of LF Refining Process ················· 122

2.4.7　Limitations of LF Refining Process ················· 125

Task 2.5　LF Refining Equipment ················· 127

2.5.1　Equipment Composition of LF Furnace ················· 128

2.5.2　LF Furnace Electrode ················· 138

2.5.3　LF Refining Equipment Improvement ················· 140

Task 2.6　ASEA-SKF Ladle Refining ················· 143

2.6.1　ASEA-SKF Ladle Refining Furnace Process Operation ················· 143

2.6.2　ASEA-SKF Ladle Refining Furnace Equipment Structure ……………… 144

2.6.3　Refining Effect of ASEA-SKF Ladle Refining Furnace ………………… 147

2.6.4　Evaluation of ASEA-SKF Ladle Refining Furnace ……………………… 148

Task 2.7　CAS Refining …………………………………………………………… 150

2.7.1　CAS Process Flow ………………………………………………………… 150

2.7.2　CAS Refining Equipment ………………………………………………… 151

2.7.3　CAS Refining Function and Effect ……………………………………… 152

2.7.4　CAS-OB Refining ………………………………………………………… 153

2.7.5　IR-UT ……………………………………………………………………… 167

Task 2.8　VD Refining …………………………………………………………… 171

2.8.1　Production and Progress of VD Refining ……………………………… 172

2.8.2　Device of VD Refining …………………………………………………… 175

2.8.3　VD Refining Limitations ………………………………………………… 176

2.8.4　Combination of VD and LF——LFV …………………………………… 176

Task 2.9　VAD Refining and VOD Refining ………………………………… 181

2.9.1　VAD Process Characteristics …………………………………………… 181

2.9.2　The Main VAD Equipment ……………………………………………… 183

2.9.3　Refining Effect …………………………………………………………… 184

2.9.4　VOD Refining Process Characteristics ………………………………… 184

2.9.5　VOD Refining Process …………………………………………………… 185

2.9.6　VOD Refining Equipment ………………………………………………… 186

Task 2.10　DH Refining ………………………………………………………… 187

2.10.1　The Equipment of DH Refining ………………………………………… 188

2.10.2　DH Refining Process …………………………………………………… 188

Task 2.11　RH Refining ………………………………………………………… 192

2.11.1　RH Refining Principle and Characteristics …………………………… 193

2.11.2　RH Refining Equipment ………………………………………………… 195

2.11.3　RH Refining Process RH ……………………………………………… 196

2.11.4　RH Refining Vacuum Treatment ……………………………………… 198

Task 2.12　AOD Refining ……………………………………………………… 200

2.12.1　Refining Equipment of AOD …………………………………………… 201

2.12.2　Principle of AOD ………………………………………………………… 202

2.12.3 AOD Process ··········· 203

Project 3 Basic Knowledge of Refractories for Secondary Refining ········ 206

Task 3.1 Requirements of Refractories for Secondary Refining ············ 206

Task 3.2 Varieties and Types of Refractory Materials for External Refining ······ 207

References ··········· 210

目 录

0 绪论 ... 1
0.1 炉外精炼技术发展的原因 ... 1
0.2 炉外精炼技术的发展历程 ... 3
0.2.1 真空处理 ... 4
0.2.2 非真空处理 ... 5
0.2.3 中国炉外精炼技术的发展 ... 8
0.2.4 炉外精炼在现代钢铁生产中的作用及特点 ... 10
0.2.5 炉外精炼的主要目的和任务 ... 14
0.2.6 精炼设备的选择问题 ... 15
0.2.7 炉外精炼技术的发展趋势 ... 17

项目1 炉外精炼手段与工艺选择 ... 19
任务1.1 合成渣洗 ... 21
1.1.1 合成渣的物理化学性能 ... 22
1.1.2 渣洗工艺 ... 27
1.1.3 应用技术 ... 30
1.1.4 作用效果及不足 ... 31
任务1.2 搅拌 ... 34
1.2.1 搅拌方法 ... 35
1.2.2 搅拌与混匀作用 ... 43
任务1.3 加热 ... 43
1.3.1 加热方法 ... 44
1.3.2 加热工艺的选择 ... 51

任务 1.4	真空	52
1.4.1	真空及其在冶金技术中的利用	54
1.4.2	钢液的真空脱气	56
1.4.3	钢液的真空脱氧	63
任务 1.5	喷吹	67

项目 2 炉外精炼基本工艺 … 71

任务 2.1	**LF 工艺**	74
2.1.1	LF 生产工艺过程	76
2.1.2	主要工艺制度	81
任务 2.2	**LF 精炼造渣技术**	88
2.2.1	合成渣的性质	90
2.2.2	夹杂物的去除	92
2.2.3	脱硫工艺	93
任务 2.3	**LF 泡沫渣精炼工艺**	103
2.3.1	泡沫渣的冶金作用	104
2.3.2	炉渣的发泡	104
2.3.3	LF 埋弧精炼渣的发泡剂	109
2.3.4	操作工艺的影响	114
2.3.5	泡沫渣的探测和控制技术	116
任务 2.4	**典型钢种 LF 精炼工艺**	117
2.4.1	普碳和低合金精炼工艺	117
2.4.2	中高碳钢精炼工艺	118
2.4.3	优质板材精炼工艺	119
2.4.4	低碳含铝钢精炼用渣	120
2.4.5	轴承钢的冶炼	121
2.4.6	其他 LF 精炼工艺新技术	122
2.4.7	LF 炉精炼工艺的局限性	125
任务 2.5	**LF 炉精炼设备**	127
2.5.1	LF 炉的设备组成	128
2.5.2	LF 炉电极	138
2.5.3	LF 精炼的设备改进	140

任务 2.6　ASEA-SKF 钢包精炼 ·· 143
2.6.1　ASEA-SKF 钢包精炼炉工艺操作 ······································ 143
2.6.2　ASEA-SKF 钢包精炼炉设备结构 ······································ 144
2.6.3　ASEA-SKF 钢包精炼炉精炼效果 ······································ 147
2.6.4　ASEA-SKF 钢包精炼炉的使用评价 ···································· 148

任务 2.7　CAS 精炼 ·· 150
2.7.1　CAS 工艺流程 ··· 150
2.7.2　CAS 精炼设备 ··· 151
2.7.3　CAS 精炼功能及效果 ·· 152
2.7.4　CAS-OB 精炼 ·· 153
2.7.5　IR-UT ·· 167

任务 2.8　VD 精炼 ·· 171
2.8.1　VD 精炼工艺及流程 ··· 172
2.8.2　VD 精炼设备 ··· 175
2.8.3　VD 精炼局限性 ··· 176
2.8.4　VD 与 LF 结合——LFV 精炼 ·· 176

任务 2.9　VAD 精炼与 VOD 精炼 ··· 181
2.9.1　VAD 工艺特点 ··· 181
2.9.2　VAD 主要设备 ··· 183
2.9.3　精炼效果 ·· 184
2.9.4　VOD 精炼工艺特点 ·· 184
2.9.5　VOD 精炼过程 ··· 185
2.9.6　VOD 精炼设备 ··· 186

任务 2.10　DH 精炼 ·· 187
2.10.1　DH 精炼设备 ·· 188
2.10.2　DH 精炼工艺 ·· 188

任务 2.11　RH 精炼 ·· 192
2.11.1　RH 精炼原理及特点 ·· 193
2.11.2　RH 精炼设备 ·· 195
2.11.3　RH 精炼工艺过程 ·· 196
2.11.4　RH 精炼真空处理 ·· 198

任务 2.12　AOD 精炼	200
2.12.1　AOD 精炼设备	201
2.12.2　AOD 法工作原理	202
2.12.3　AOD 法工艺过程	203

项目 3　炉外精炼用耐火材料基础知识 …… 206

任务 3.1　炉外精炼对耐火材料的要求	206
任务 3.2　炉外精炼用耐材的品种与类型	207

参考文献 …… 210

0 Introduction
0 绪论

Secondary refining refers to the steelmaking process in which the initial molten steel in converter or electric furnace is poured into ladle or special vessel for deoxidization, desulfurization, decarburization, degassing, removal of non-metallic inclusions, adjustment of liquid steel composition and temperature to achieve further smelting purpose. The refining task to be completed in conventional steelmaking furnace, such as removal of impurities (including unnecessary elements, gases and inclusions). The task of adjusting and homogenizing the composition and temperature is carried out in the ladle or other containers in part or in whole. The one-step steel-making method is changed into two-step steel-making, which is divided into two steps of primary refining and refining is also known as: Secondary refining, Secondary steelmaking, Secondary metallurgy, and Ladle metallurgy.

炉外精炼是指将在转炉或电炉内初炼的钢液倒入钢包或专用容器内进行脱氧、脱硫、脱碳、去气、去除非金属夹杂物和调整钢液成分和温度，以达到进一步冶炼目的的炼钢工艺（即将在常规炼钢炉中完成的精炼任务）。如去除杂质（包括不需要的元素、气体和夹杂），调整和均匀成分和温度（在钢包或其他容器中进行），变一步炼钢法为二步炼钢（把传统的炼钢过程分为初炼和精炼两步进行，也称为二次精炼、二次炼钢、二次冶金和钢包冶金）。

0.1 The Reasons for Development of Secondary Refining
0.1 炉外精炼技术发展的原因

Secondary Refining is to degass, deoxidize, desulfurize, remove inclusions and fine tune the composition of the primary molten steel in the container of vacuum, inert gas or reducing atmosphere. Secondary Refining can improve the quality of steel, shorten smelting time, optimize process and reduce production cost.

炉外精炼是将初炼的钢液在真空、惰性气体或还原性气氛的容器内进行脱气、脱氧、脱硫、去除夹杂物和成分微调等操作。炉外精炼可提高钢的质量，缩短冶炼时间，优化工艺过程，并降低生产成本。

At first, secondary refining was limited to the production of special steel and high-quality steel, and then expanded to the production of ordinary steel. Now it has basically become an essential link in the steel-making process. It is a bridge connecting smelting and continuous casting, which is used to coordinate the normal production of steel-making and continuous casting. In the future, iron and steel production will develop towards the direction of high integration of near final continuous casting (such as thin slab) and subsequent processes, which requires that the cast billets have no defects and can realize high continuous operation in operation. Therefore, in order to improve the quality of molten steel, it is necessary to further develop the refining technology outside the furnace, so that the smelting, casting and rolling processes can achieve the best connection, so as to improve productivity and improve product quality, and reduce production costs.

炉外精炼起初仅限于生产特殊钢和优质钢,后来扩大到普通钢的生产上,现在已基本成为炼钢工艺中必不可少的环节。炉外精炼是连接冶炼与连铸的桥梁,用以协调炼钢和连铸的正常生产。未来的钢铁生产将向着近终型连铸(如薄板坯)和后步工序高度一体化的方向发展,这就要求浇铸出的铸坯无缺陷,并且能在操作上实现高度连续化作业。因此,若要求钢液具有更高的质量特性,那就必须进一步发展炉外精炼技术,使冶炼、浇铸和轧制等工序实现最佳衔接,进而达到提高生产率、降低生产成本、提高产品质量的目的。

With the progress of science, it is required to improve the quality of steel. For example, in terms of transportation, with the development of high-speed trains, light vehicles and large ships, the purity of steel used needs to be continuously improved; in terms of petrochemical industry, in terms of deep oil production, long-distance high-pressure oil and gas transportation, the strength and performance of steel used need to be continuously improved. The development of iron and steel industry is no longer the pursuit of output, but also the need to improve the purity of steel and quality uniformity, and enhance the strength, accuracy and performance of products. One of the key points to improve the performance and quality of steel is the ultra pure purification of steel, which effectively reduces the content of harmful impurity elements and inclusions in steel, greatly reduces the tendency of defects such as center segregation, cracks, large inclusions, pores, white spots and spots, makes the structure of steel compact and uniform, and improves the surface and internal quality of continuous casting slab and the performance of steel. Precise control of chemical composition can ensure the stability of steel performance. By reducing the content of P and S in the steel, the impact performance, lamellar cracking resistance and thermal brittleness can be improved, the center segregation can be reduced, and the surface defects of continuous casting slab can be prevented. By reducing the content of oxygen, hydrogen and nitrogen in the steel, the defects of ultrasonic detection and strip cracks can be reduced, and the pipe making performance of the steel can be improved. The shape of inclusions can be controlled. To improve the deep drawing property and processing property of steel, it is important for various refining technologies outside the furnace to obtain these high-quality steel.

随着科学的进步,钢材质量的要求日益提高。例如,在交通运输方面,随着高速火车、汽车的轻型化高速化,船舶的大型化,所用钢材的纯净度需要不断提高;在石油化工方面,深层采油和长距离高压输油输气,也需要不断提高所用钢材的强度和性能。钢铁产业的发展已不再是追求产量,而是需要提高钢材的纯净度、质量的均匀性,增强产品的强度、精度和性能。提高钢材性能和质量的关键是钢的超纯净化,即有效降低钢中有害杂质元素和夹杂物的含量,从而大大减轻中心偏析、裂纹、大型夹杂、气孔、白点和斑疤等缺陷的产生,使钢组织致密均匀,改善连铸坯表面及内部质量,使钢材性能大幅度提高。同时也需要精确控制化学成分,从而保证钢性能的稳定,通过减少钢中P、S含量,来改善冲击性能、抗层状拉裂性能和热脆性,减少中心偏析和防止连铸坯的表面缺陷;减少钢中O、H、N含量,以减少超声波探伤缺陷和条状裂纹等,改善钢材的制管性能;控制夹杂物的形态来改善钢的深冲性能和钢的加工性能。各种炉外精炼技术则是获得这些高质量钢材的重要措施。

With the progress of modern science and technology, and the development of industry, the requirement of steel quality (such as the purity of steel) is becoming higher and higher. The molten steel produced by common steelmaking furnace (converter, electric furnace and flat furnace) has been unable to meet the requirement of its quality. In order to improve the productivity and shorten the smelting time, part of the task of steelmaking is also expected to be completed outside the furnace. In addition, the development of continuous casting technology also puts forward strict requirements for the composition, temperature and gas content of the molten steel.

随着现代科学技术的进步和工业的发展,钢质量(如钢的纯净度)的要求越来越高,用普通炼钢炉(转炉、电炉和平炉)冶炼出来的钢液已经难以满足其质量的要求。为了提高生产率,缩短冶炼时间,也希望能把炼钢的一部分任务移到炉外去完成。另外,连铸技术的发展对钢液的成分、温度和气体的含量等也提出了严格的要求。

In a word, the use of secondary refining can improve the quality of steel, expand varieties, shorten smelting time, improve productivity and economic benefits, adjust the production rhythm of steelmaking furnace and continuous casting, and reduce the cost of steelmaking.

总之,采用炉外处理技术可以提高钢的质量,扩大品种,缩短冶炼时间,提高生产率,调节炼钢炉与连铸的生产节奏,并可降低炼钢成本,提高经济效益。

0.2 Development History of Secondary Refining
0.2 炉外精炼技术的发展历程

In the 1930s~1940s, high basicity synthetic slag was used to remove S in order to improve the quality of molten steel, and vacuum mold casting technology was used to degasify molten steel under low vacuum. In 1933, R. Perrin, France, applied high basicity synthetic slag to carry out 'Slag Washing and Desulfuration' for molten steel——The rudiment of modern refining technology outside the furnace.

20世纪30~40年代,使用高碱度合成渣进行脱S,从而提高钢水质量;真空模铸技术在低真空度下对钢水进行脱气处理。1933年,法国佩兰(R. Perrin)应用高碱度合成渣,对钢液进行"渣洗脱硫"——现代炉外精炼技术的萌芽。

In 1950, the Gazal method (France) was born to mix molten steel with argon blowing at the bottom of ladle to uniform its composition, temperature and remove inclusions.

1950年,Gazal法(法国)诞生。该法将钢水与钢包底部吹氩混合,使钢水成分、温度均匀，并去除夹杂物。

In the 1950s, the development of vacuum technology and the successful development of large-scale steam jet pump provided conditions for large-scale vacuum treatment of liquid steel. Various vacuum treatment methods of molten steel, degassing method of ladle and reverse ladle treatment method (BV method) have been developed. And typical methods include DH method developed by Dortmund and Horder before 1957, and RH method jointly invented by Ruhrstahl and Heraeus before 1957.

20 世纪 50 年代，真空技术的发展和大型蒸汽喷射泵的研制成功，为钢液的大规模真空处理提供条件，并开发出了各种钢液真空处理方法，如钢包除气法、倒包处理法（BV 法）等。典型方法包括：1957 年西德的多特蒙德（Dortmund）和豪特尔（Horder）两公司开发的提升脱气法（DH 法），1957 年西德的德鲁尔钢铁公司（Ruhrstahl）和海拉斯公司（Heraeus）共同发明的真空循环脱气法（RH 法）等。

In the 1960s and 1970s, the invention of refining methods was flourishing, which was closely related to the production of clean steel, the requirement of stable composition and temperature of molten steel, and the expansion of steel varieties. In this period, on the basis of vacuum degassing, VOD, VaD, ASEA, LFV and RH-OB were developed vigorously, and two series of vacuum and non vacuum refining technologies were formed.

20 世纪 60~70 年代是炉外精炼方法发明的繁荣时期。这与该时期提出洁净钢生产，连铸要求稳定的钢水成分和温度，以及扩大钢的品种密切相关联。在这个时期的真空脱气基础上，VOD、VAD、ASEA、LFV、RH-OB 蓬勃发展，炉外精炼技术形成了真空和非真空两大系列。

0.2.1　Vacuum Treatment

0.2.1　真空处理

The methods of vacuum treatment are as follows:

真空处理方法包括：

(1) Ladle degassing, gradually used in West Germany and Soviet Union in 1940s.

(1) 钢包除气法：20 世纪 40 年代，该方法在西德和苏联逐步使用。

(2) Rewinding (now eliminated), such as BV (used in West Germany in 1952).

(2) 倒包处理法（现已被淘汰），如 BV 法（1952 年西德使用）。

(3) Degassing in tapping process (TD method, used in West Germany in 1962).

(3) 出钢过程脱气法（TD 法），1962 年该方法在西德使用。

(4) Vacuum pouring method, first used in Soviet Union in 1950s.

(4) 真空浇注法，前苏联是第一个使用该方法的国家。

(5) Vacuum argon blowing method (Finkl or Gazid method), used in France and Britain from 1953 to 1963).

(5) 真空吹氩法（Finkl 或 Gazid 法），1953~1963 年，该方法在法国、英国使用。

(6) The vacuum oxygen decarburization method (VOD), developed in 1965 in West Germa-

ny for the production of ultra-low carbon stainless steel, and the vacuum arc heating decarburization method (VAD) in the United States in 1967.

(6) 真空吹氧脱碳法(VOD)，1965年发明于德国，用于超低碳不锈钢生产；真空电弧加热去气法(VAD)，1967年发明于美国。

(7) ASEA-SKF, developed in Sweden for the production of stainless steel and bearing steel in 1965, including arc heating, electromagnetic stirring and vacuum degassing;

(7) 钢包精炼炉法(ASEA-SKF)，1965年发明于瑞典，用于不锈钢和轴承钢生产的有电弧加热、电磁搅拌和真空脱气。

(8) RH-OB, developed in Japan in 1978, to improve the production efficiency of ultra-low carbon steel.

(8) RH吹氧法(RH-OB)，1978年发明于日本，用于提高超低碳钢生产效率。

0.2.2 Non Vacuum Treatment

0.2.2 非真空处理

The methods of non vacuum treatment are as follows:

非真空处理方法包括：

(1) The argon oxygen decarburization refining method (AOD) developed in the United States in 1968 for the production of low carbon stainless steel.

(1) 1968年美国发明的用于低碳不锈钢生产的氩氧脱碳精炼法(AOD)。

(2) The ladle furnace method (LF) developed in Japan with ultra-high power electric arc furnace to replace the reduction period of EAF in 1971. LF—VD was developed with vacuum degassing technology later.

(2) 1971年日本开发的钢包炉法(LF)，代替电炉还原期对钢水进行精炼。LF法后来配套真空脱气技术，发展形成LF—VD。

(3) Spray metallurgy technology, such as SL method, developed in Sweden in 1976, TN method developed in West Germany before 1974 and Kip method developed in Japan.

(3) 喷射冶金技术，如1976年瑞典开发的氏兰法(SL法)，1974年前西德开发的蒂森法(TN法)，以及日本开发的川崎喷粉法(KIP)。

(4) Alloy cored wire feeding technology, such as wire feeding method (WF) developed in Japan in 1976.

(4) 喂合金包芯线技术，如1976年日本开发的喂丝法(WF)。

(5) The argon blowing technology with cover or slag cover, such as the sealed argon blowing method (SAB method) and ladle argon blowing method (CAB method) developed in Japan in 1965, and the composition adjusted sealed argon blowing method (CAS method) developed in Japan in 1975.

(5) 加盖或加浸渣罩的吹氩技术，如1965年日本发明的密封吹氩法(SAB法)和带盖钢包吹氩法(CAB法)，以及1975年日本发明的成分调整密封吹氩法(CAS法)。

The technology developed in this period also includes VOD—VAD, ASEA—SKF, RH-OB,

LF, jet metallurgy technology (SL, TN, KTS, KIP), alloy cored wire technology, and argon blowing technology (SAB, CAB, CAS) with cover and impregnation cover. This period basically established the position and role of argon blowing technology as the basis of various secondary refining technologies.

这一时期发展的技术还包括 VOD—VAD、ASEA—SKF、RH—OB、LF、喷射冶金技术（SL、TN、KTS、KIP）、合金包芯线技术和加盖和加浸渍罩的吹氩技术（SAB、CAB、CAS）。这个时期基本奠定了吹氩技术在各种炉外精炼技术基础的地位和作用。

Since the 1980s, there has been an alloy clad wire feeding method and the combination of refining equipment with different functions. The secondary refining has become a symbol of the level of modern iron and steel production process, high quality of iron and steel products. And the secondary is also refining developing and improving in the direction of more complete functions, higher efficiency and better metallurgical effect. The technologies developed in this period mainly include RH-KTB, RH-MFB, RH-IJ, RH-PB, V-KIP, and IR-UT.

20 世纪 80 年代以来，出现了合金包丝线喂丝法、不同功能精炼设备的使用组合。炉外精炼已成为现代钢铁生产流程水平和钢铁产品高质量的标志，同时正朝着功能更全、效率更高、冶金效果更佳的方向发展和完善。这一时期发展起来的技术主要包括 RH 顶吹氧法、RH 多功能氧枪、RH 钢包喷粉法、RH 真空室喷粉法、真空川崎喷粉法和吹氧喷粉升温精炼法。

In the 1990s, various forms of secondary refining have been developed, which inlude circulating vacuum treatment (RH, RH-OB), lifting vacuum treatment (DH), alloy fine adjustment and temperature treatment (CAS, CAS-OB), LF ladle refining furnace, vacuum argon deoxidization (VD), vacuum oxygen decarburization (VOD), ladle powder injection (kip, TN, SL), wire feeding, argon oxygen refining (AOD), etc.

20 世纪 90 年代，炉外精炼技术以多种形式发展，广泛应用的技术包括循环真空处理（RH, RH-OB）、提升真空处理（DH）、合金微调及温度处理（CAS, CAS-OB）、LF 钢包精炼炉、真空吹氩脱氧（VD）、真空吹氧脱碳（VOD）、钢包喷粉（KIP, TN, SL）、喂丝法和氩氧精炼（AOD）等。

With the development of continuous casting, the quality requirements of continuous casting slab and the connection between steelmaking furnace and continuous casting promote rapidly, with the continuous requirements of increasing production efficiency and shortening production cycle. The secondary refining is developing towards the direction of high efficiency and large-scale. Since the 21st century, some fine steel grades such as pure steel have been put into using one after another. With higher rhythm, rich means and super strong application ability, the secondary refining has become an indispensable process in the steel production process. The summary of development history of secondary refining is shown in Table 0-1.

随着连铸的发展，连铸坯对质量的要求及炼钢炉与连铸的衔接日益提高，同时不断要求增加生产效率，缩短生产周期。炉外精炼技术向着高效率、大型化方向发展。21 世纪以来，纯净钢等一些精品钢种相继投入使用，炉外精炼技术以更高节奏、丰富的手段、超强的应用能力成为钢铁生产过程不可缺少的工序。炉外精炼技术发展历程见表 0-1。

Table 0-1 Summary of development history of secondary refining
表 0-1 炉外精炼技术发展历程部分汇总表

Name 名称	Time 开发年代	Developer 国家	Technical characteristics 技术特点
Synthetic slag washing 合成渣渣洗	1930s	France 法国	Liquid slag mixing, 60%CaO—40%Al_2O_3 (by mass) 液渣冲混，60%CaO—40%Al_2O_3（质量分数）
SAB, CAB, CAS	1974	Japan 日本	Ladle Cap (cover), Ar blowing, alloy adding 钢包加盖（罩），吹 Ar，加合金
CAS-OB	1983	Japan 日本	Temperature rise of $Al-O_2$ reaction based on CAS CAS 基础上的 $Al-O_2$ 反应提温
VID	1950s	Germany 德国	Vacuum degassing of ladle (steel flow, tapping) 钢包（钢流、出钢）真空脱气
DH	1956	Germany 德国	Lifting vacuum degassing 提升式真空脱气
RH	1959	Germany 德国	Circulating vacuum degassing 循环式真空脱气
RH-OB	1972	Japan 日本	Oxygen blowing under RH vacuum tank submerged RH 真空槽下部淹没吹氧
RH-KTB	1988	Japan 日本	Top blowing oxygen in RH vacuum tank RH 真空槽内顶吹氧
VOD	1965	Germany 德国	Oxygen blowing under vacuum 真空下顶吹氧
AOD	1968	America 美国	$Ar-O_2$ Submerged injection $Ar-O_2$ 淹没喷吹
VAD	1968	America 美国	Arc heating, Ar blowing, and ladle degassing under low pressure 低压下电弧加热，吹 Ar，钢包脱气
ASEA-SKF	1965	Sweden 瑞典	Electromagnetic stirring, arc heating under atmosphere, and degassing of ladle 电磁搅拌，大气下电弧加热，钢包脱气
LF	1971	Japan 日本	Arc heating under atmosphere, and bottom blowing Ar 大气下电弧加热，底吹 Ar
Dusting	1970s	Germany 德国	Submerged blowing slag powder or alloy powder 淹没喷吹渣粉或合金粉
Feed Wire	1970s	France 法国	High speed feeding core wire (alloy material) 高速喂入包芯线（合金料）
SRP	1982	Japan 日本	Pretreatment of converter duplex—slag gold countercurrent hot metal 转炉双联—渣金逆流铁水预处理
ORP	1986	Japan 日本	Pretreatment of hot metal ladle with immersion hood as reactor 带浸渍罩的铁水包为反应器的预处理

0.2.3 Development of China's Secondary Refining

0.2.3 中国炉外精炼技术的发展

Since the founding of new China, China has continuously increased the construction of iron and steel industry, and increased investment in science and technology. With the continuous improvement of the quality requirements of steel products, the refining technology outside the furnace is developing rapidly.

新中国成立,我国不断增加钢铁产业建设,加大科技投入。随着对钢铁产品质量要求的不断提高,炉外精炼技术不断发展。

In the 1960s~1970s, special steel enterprises, mechanical and electrical industries, and military industries applied molten steel refining technology. The refining technologies of Daye and WISCO, Beijing heavy-duty ASEA-SKF, Fuzhou steel VOD-VAD and Shougang ladle argon blowing have been put into use successively. In the 1980s, the development of off furnace treatment technology laid the foundation. LF furnace, alloy cored wire and wire feeding equipment, hot metal injection desulfurization and injection metallurgy technology have been gradually localized. But the complete set of refining equipment in large-scale iron and steel enterprises still need to be imported. For example, Baosteel introduces large-scale RH device and Kip powder spraying device; Shagang introduces SL technology; Baosteel and TISCO introduce hot metal triple removal technology and equipment.

20世纪60~70年代,特钢企业和机电、军工行业应用钢水精炼技术。大冶、武钢的RH,北京重型的ASEA-SKF,抚钢的VOD-VAD,以及首钢的钢包吹氩等炉外精炼技术相继投入使用。20世纪80年代是我国炉外处理技术发展奠定基础的时期。LF炉、合金包芯线及喂线设备、铁水喷射脱硫和喷射冶金技术已逐步趋于国产化,但大型钢铁企业的成套精炼设备还需要进口。例如:宝钢引进大型RH装置和KIP喷粉装置;沙钢引进SL技术;宝钢、太钢引进铁水三脱技术与装备。

In the 1990s, the technology of out of furnace treatment in China has developed rapidly. With the gradual improvement of the quality of thin strip products, the strength and performance of rod and wire continue to be optimized. LF furnace, VD vacuum technology, and RH system have been widely used in major iron and steel production enterprises. At the first National Conference on off furnace treatment technology in 1991, the basic policy of 'based on products, reasonable selection, system matching, and online' for the development of off furnace treatment technology was defined, and four guiding principles for the development of off furnace treatment technology were defined as follows:

20世纪90年代,我国的炉外处理技术得到迅速发展,随着薄带产品质量逐步提高,棒线材强度、性能不断优化,LF炉、VD真空技术、RH系统等已经广泛应用于各大钢铁生产企业。1991年在全国首次炉外处理技术工作会议上,明确了"立足产品、合理选择、系统配套、强调在线"的发展炉外处理技术的基本方针,明确了四条发展炉外处理技术的指导思想,其分别为:

(1) It is an important link of modern metallurgical production to make clear that the treatment outside the furnace is an important strategic measure to promote the technological progress of iron and steel industry as a whole.

(1) 明确炉外处理是现代冶金生产的一个重要环节,是从整体上推动钢铁工业技术进步的重大战略措施。

(2) In order to develop the off furnace treatment technology, the methods with corresponding functions must be selected according to the requirements of steel grades and users for product quality, in combination with the raw materials and process characteristics of the enterprise. The basic process ideas must be clear. When promoting the off furnace treatment technology, it is advocated to not only create corresponding conditions for positive development, but also prevent the rush to success.

(2) 发展炉外处理技术,必须根据钢种和用户对产品质量的要求,结合本企业的原料和工艺特点,选择具有相应功能的方法,基本工艺思路必须明确。在推进炉外处理技术时,既要创造相应条件积极发展,又要防止急于求成仓促上马。

(3) Fully consider the current and future needs of continuous casting production development to ensure the connection of steelmaking and continuous casting production, and the improvement of production efficiency.

(3) 充分考虑当前和将来连铸生产发展的需要,保证炼钢—连铸生产的衔接和生产效率的提高。

(4) It is a system engineering idea to set up the treatment outside the furnace. It is necessary to ensure that the main body and auxiliary equipment are complete according to the requirements of the system engineering, pay attention to the characteristics of different enterprises and the requirements of different levels, and ensure the proper level of technology and equipment and the proper depth of investment.

(4) 树立炉外处理是一项系统工程的思想,必须按系统工程的要求确保主体和辅助设备配套齐全,并注意不同企业的特点及不同层次要求,确保应有的工艺和装备水平,确保应有的投资深度。

In the 1990s, China has made four outstanding achievements in refining technology outside the furnace:

20世纪90年代,我国取得的四项突出炉外精炼技术成果包括:

(1) Development and application of comprehensive refining technology for vacuum treatment of molten steel.

(1) 钢水真空处理综合精炼技术开发与应用。

(2) Development and application of MgO hot metal desulfurization technology and converter hot metal pretreatment technology.

(2) 镁质铁水脱硫技术和转炉铁水预处理技术开发与应用。

(3) Development and application of refining technology for medium and small ladle.

(3) 适于中小钢包钢水精炼技术的开发与生产应用的发展。

(4) The development and application of metallurgical technology in tundish centering on

Mg—Ca—Zr system material and flow field optimization are combined with the basic technology of argon blowing and wire feeding in ladle refining furnace.

（4）中间包以镁—钙—锆系材料及流场优化为中心的中间包冶金技术、再与钢包精炼炉吹氩、喂丝等基本技术相结合的开发与应用

In the 21st century, with the gradual development of China from a large iron and steel country to a strengthened iron and steel industry, the secondary refining has achieved comprehensive, high-level and rapid development. The construction of large-scale fine steel base, high-speed railway, shipbuilding, aviation equipment and other fields have promoted the secondary refining to mature, and the secondary refining equipment has gradually become intelligent, systematic and intensive, which not only guarantees the technology. The production of iron and steel quality, more from the effective resources and environmental protection of the development of China's iron and steel industry continue to inject vitality. Refining out of furnace has become an indispensable process in iron and steel smelting process.

21世纪随着我国逐步由钢铁大国向钢铁强化迈进，炉外精炼技术得到了全面发展，大型精品钢铁基地的建设、高速铁路、船舶制造、航空器材等领域助推炉外精炼技术走向成熟，炉外精炼设备也逐步趋于智能化、系统化、集约化。这样不仅从工艺上保证生产钢铁质量，更从资源利用、环境保护等方面为我国钢铁产业的发展不断注入活力。炉外精炼已成为钢铁冶炼过程不可缺少的工序。

0.2.4 Role and Characteristics of Refining in Modern Steel Production

0.2.4 炉外精炼在现代钢铁生产中的作用及特点

0.2.4.1 The Role and Position of Refining Outside in Modern Iron and Steel Production

0.2.4.1 炉外精炼在现代钢铁生产中的作用及地位

The traditional iron and steel production process is: (BF)Blast Furnace→(BOF)Blowing Oxygen Furnace→(CC)Continuous Casting→Rolling.

传统的高炉—氧气转炉钢铁生产工艺流程为：高炉→转炉→连铸→轧制。

In the past two decades, the international iron and steel circles think that the two most successful production processes are as follows:

(1) Blast furnace smelting from ore → Hot metal pretreatment → Converter blowing → Secondary refining→Continuous casting→Continuous rolling long process.

(2) Electric furnace smelting from scrap→Secondary refining→Continuous casting→Continuous rolling short process.

近20年来国际钢铁界认为最成功的两个生产流程是：

（1）从矿石开始的高炉冶炼→铁水预处理→转炉吹炼→二次精炼→连铸→连轧长流程；

（2）从废钢为原料的电炉熔炼→二次精炼→连铸→连轧短流程。

The technological process of iron and steel production in modern blast furnace—oxygen converter is: Blast furnace→Hot metal pretreatment→Top bottom combined blowing of oxygen converter→External refining (RH, CAS-OB, LF)→Continuous casting and rolling or continuous casting→Hot delivery and direct rolling of slab. As shown in Figure 0-1, the converter and electric furnace lose their original steelmaking functions: The converter only plays the role of decarburization and temperature rise of molten iron, and the electric furnace operation only has the melting period, mainly completing the requirements of scrap melting and temperature rise of molten steel. Although the secondary refining weakens the smelting function of converter and electric furnace, it improves the purity of steel continuously and meets the requirements of continuous casting. More importantly, it is easy to match with continuous casting and continuous rolling, shortens the smelting cycle and improves the production efficiency. Secondary refining is known as an important technology of iron and steel metallurgy which has developed rapidly in half a century. Its functions are summarized as follows:

现代高炉—氧气转炉钢铁生产工艺流程为：高炉→铁水预处理→氧气转炉顶底复合吹炼→炉外精炼（RH，CAS-OB，LF）→连铸连轧或连铸→铸坯热送和直接轧制。如图0-1所示，通过炉外精炼，转炉和电炉失去了原有炼钢功能，转炉仅起到铁水脱碳和提温的作用，电炉操作也只有熔化期，主要完成废钢熔化及钢水提温的要求。虽然炉外精炼弱化了转炉和电炉的冶炼功能，但却提高了钢的纯净度，满足连铸要求；容易与连铸—连轧匹配，缩短冶炼周期，提高生产效率。炉外精炼被誉为半个世纪以来迅速发展的钢铁冶金重要技术，其作用归纳为：

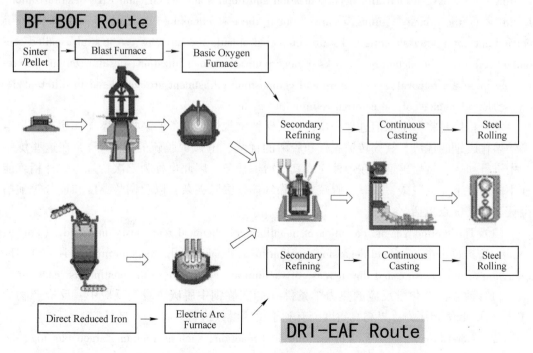

Figure 0-1　Process position of refining outside in modern iron and steel production process
图 0-1　现代钢铁生产工艺流程图

（1）It is an indispensable means to improve the quality of metallurgical products and expand the variety of iron and steel production.

（1）炉外精炼是提高冶金产品质量，扩大钢铁生产品种不可缺少的手段。

（2）It is a powerful means to optimize metallurgical production process, further improve production efficiency, save energy and consumption, and reduce production cost.

（2）炉外精炼是优化冶金生产工艺流程，进一步提高生产效率、节能强耗、降低生产成本的有力手段。

（3）It is the necessary technological to ensure the optimization of high temperature connection between Steel making—Continuous casting—Continuous casting.

（3）炉外精炼是保证炼钢—连铸—连铸坯热送热装和直接轧制高温连接优化的必要工艺手段。

（4）It is the independent and irreplaceable production process in the optimized and reorganized iron and steel production process.

（4）炉外精炼是优化重组的钢铁生产工艺流程中独立的，不可替代的生产工序。

0.2.4.2 Technical Characteristics of External Refining

0.2.4.2 炉外精炼的技术特点

Furnace refining, the whole steelmaking process carried out step by step, retains some advantages of conventional steelmaking equipment, such as the advantages of ultra-high power EAF melting scrap, oxygen converter decarbonization and dephosphorization, and EAF desulfurization. In this way, the primary refining is carried out in these steelmaking furnaces, then the tapping is carried out, and other smelting tasks are completed outside the furnaces, so it is also called secondary refining. The metallurgical tasks of decarbonization, deoxidization, desulfurization, gas removal, inclusion removal, temperature and composition adjustment are completed in different degrees. The advantages of the external refining are as follows:

炉外精炼是将整个炼钢过程分步进行，保留常规炼钢设备的某些优势（如超高功率电弧炉熔化废钢的优势、氧气转炉脱碳和脱磷的优势、电弧炉脱硫的优势等），在这些炼钢炉中进行初炼，然后出钢，在炉外完成其他冶炼任务，因此也称为二次精炼。炉外精炼能在不同程度上完成脱碳、脱氧、脱硫，去除气体，去除夹杂，以及调整温度和成分等项冶金任务。其优点包括：

（1）The thermodynamic condition of metallurgical chemical reaction is improved. (For example, in decarburization and degassing reaction in steelmaking, the reaction product is gas). The reduction of gas pressure and the increase of vacuum are conducive to the continuous reaction.

（1）改善冶金化学反应的热力学条件（如在炼钢中脱碳、脱气反应中，反应产物为气体），降低气相压力，提高真空度，有利于反应继续进行。

（2）Good kinetic conditions of metallurgical reaction, such as vacuum, argon blowing, degassing and powder spraying, are created. And various agitations are used to increase the mass transfer coefficient and expand the reaction interface.

（2）创造良好的冶金反应的动力学条件（如真空、吹氩、脱气、喷粉），应用各种搅拌增大传质系数，扩大反应界面。

In the process of steelmaking, most of the metallurgical reactions are multiphase reactions at high temperature. The chemical reactions themselves are usually fast, and the rate of material transfer to the reaction zone is often slow. Therefore, in most cases, the metallurgical reaction rate is limited by the material transfer rate. According to Fick's law, the mass transfer rate is directly proportional to the mass transfer coefficient, the concentration gradient and the cross-sectional area of the diffusion flow. All kinds of secondary refining technologies are from different aspects and angles to create the best possible dynamic conditions to improve the metallurgical reaction rate in different ways. For example, vacuum or argon blowing is used to increase the concentration gradient of gas between gas phase and liquid steel; powder spraying is used to increase the reaction area; and various agitation methods are used to increase the mass transfer coefficient and expand the reaction area, so as to solve the problem that the advantages of traditional steel-making equipment cannot be fully exerted.

炼钢过程中的各种冶金反应多数是在高温下进行的多相反应，通常化学反应本身进行较快，而向反应区传递物质的速率往往缓慢。因此，在多数情况下，冶金反应速率受物质的传递速率所限制。根据菲克定律，物质的传递速率与传质系数、传递物质的浓度梯度和扩散流的截面积成正比。各种炉外精炼技术都是从各个不同的方面不同的角度，以不同的方式创造尽可能好的动力学条件来改善冶金反应速率。例如，应用真空或吹氩来提高气体在气相与钢液之间的浓度梯度；应用喷粉来增加反应界面积；应用各种方式的搅拌来增大传质系数和扩大反应界面积，从而解决传统炼钢设备优越性不能充分发挥的问题。

（3）Increase the reaction area of slag steel. Various stirring or powder spraying processes are adopted for various refining equipment outside the furnace, result in steel slag emulsification, particle bubble floatation, collision, polymerization and other phenomena, and significantly increas the reaction area of slag steel and the reaction speed.

（3）增大渣钢反应面积。对各种炉外精炼设备均采用各种搅拌或喷粉工艺，造成钢渣乳化、颗粒气泡上浮、碰撞、聚合等现象，显著增加渣钢反应面积，提高反应速度。

（4）The reaction conditions should be controlled accurately and the composition and temperature of molten steel should be uniform. Most of the secondary refining equipments are equipped with various heating functions, which can accurately control the reaction temperature. At the same time, the composition of the molten steel can be adjusted precisely by stirring evenly to realize the fine adjustment of the composition. The precise control of chemical reaction conditions makes all kinds of metallurgical reactions closer to equilibrium.

（4）精确控制反应条件，均匀钢水成分和温度。多数炉外精炼设备配备了各种不同的加热功能，可以精确控制反应温度；同时，通过搅拌均匀钢水成分，精确调整成分，实现成分微调；精确控制化学反应条件，使各种冶金反应更趋近平衡。

（5）The atmosphere can be selected and controlled to meet the diversified requirements of steel-making production. External refining can cooperate with electric arc furnace and converter. The initial external refining is to cooperate with electric arc furnace to produce special steel. It is also used in the production of ordinary carbon steel. Now it has become an indispensable link in the steel-making process, especially with ultra-high power electric arc furnace, which can produce more advantages of ultra-high power technology and improve the power utilization rate of ultra-high power electric arc furnace. For example, in the production process of smelting high-quality rod and wire rod, the reduction condition of LF furnace is used for desulfurization after the initial smelting of molten steel in converter or electric arc furnace, then the oxidation condition of RH furnace is used for decarburization, and finally the casting is carried out in the continuous casting workshop, which not only has stable production rhythm and greatly shortened smelting time, but also ensures the quality of molten steel in the whole smelting process.

（5）气氛可选可控，满足炼钢生产多样化条件要求。炉外精炼可以与电弧炉和转炉配合，最初的炉外精炼是与电弧炉配合生产特殊钢，后来也运用在普碳钢的生产上，现已成为炼钢工艺中不可缺少的一个环节，尤其与超高功率电弧炉配合，更能发挥超高功率技术的优越性，从而提高超高功率电弧炉的功率利用率。例如，在冶炼优质棒线材生产过程中，转炉或电弧炉初炼钢水后，利用 LF 炉的还原性条件脱硫，再利用 RH 炉的氧化性条件脱碳，最后进入连铸车间进行浇注。这不仅能稳定生产节奏，缩短冶炼时间，同时也保证钢水在冶炼全程的质量。

（6）Improve on-line detection facilities and realize computer intelligent control of refining process. Ensure the hit rate and control accuracy of refining end point, and improve the stability of product quality.

（6）健全在线检测设施，对精炼过程实现计算机智能化控制。保证精炼终点的命中率和控制精度，提高产品质量的稳定性。

0.2.5 The Main Purpose and Task of Secondary Refining

0.2.5 炉外精炼的主要目的和任务

In the aspect of improving the quality of steel, Steelmaking technology is always in the direction of reducing the content of harmful impurities and non-metallic inclusions in steel, improving the shape and distribution of inclusions, making the chemical composition of steel uniform, accurately controlling the process temperature, and making it suitable for the production requirements of later processes. In the aspect of economy, it is in the direction of increasing productivity, reducing raw materials, energy and labor. In the industry, it is required to improve the adaptability of multi steel production. The main purpose of secondary refining is to improve the quality of molten steel, remove harmful elements and inclusions in molten steel and purify the quality of molten steel according to metallurgical thermodynamics, metallurgical dynamics and transmission principle. The following tasks can be completed in the secondary refining:

在提高钢的质量方面，炼钢技术是向降低钢中的有害杂质和非金属夹杂物含量，改善夹杂物的形态和分布，使钢的化学成分均匀，精确控制过程温度，使之能适合后部工序生产要求的方向发展；在经济方面是向提高生产率、降低原材料、能源和劳动力消耗方向发展；在工业方面，则要求尽量提高生产多钢种的适应能力。炉外精炼主要目的是为提高钢水质量，依据冶金热力学、冶金动力学，传输原理去除钢水内部有害元素及夹杂物，从而净化钢水质量。炉外精炼可以完成下列任务：

(1) In order to improve the purity and properties of steel, the contents of oxygen, sulfur, hydrogen, nitrogen and non-metallic inclusions in steel are reduced and the shape of inclusions is changed.

(1) 降低钢中氧、硫、氢、氮和非金属夹杂物含量，改变夹杂物形态，从而提高钢的纯净度，改善钢的性能。

(2) Deep decarburization can meet the requirements of low carbon or ultra-low carbon steel.

(2) 深脱碳，满足低碳或超低碳钢的要求。

(3) The alloy composition is fined tune to make it evenly distributed, and reduce the consumption of alloy to improve the yield of alloy.

(3) 微调合金成分，使其分布均匀，降低合金的消耗，从而提高合金收得率。

(4) Adjust the temperature of molten steel to the range required by pouring to reduce the temperature gradient of molten steel in ladle.

(4) 调整钢水温度到浇注所要求的范围内，减小包内钢水的温度梯度。

The above tasks can improve the quality and the productivity, expand the variety, reduce the consumption and cost, shorten the smelting time, and coordinate the production of steelmaking and continuous casting. However, there is no one refining method outside the furnace to complete all the above tasks, and only one or several tasks can be completed by one refining method. Due to different plant conditions and smelting steel grades, generally one or two kinds of secondary refining equipment are equipped according to different needs.

完成上述任务就能达到提高质量、扩大品种、降低消耗和成本、缩短冶炼时间、提高生产率以及协调好炼钢和连铸生产的配合等目的。但是，到目前为止，还没有任何一种炉外精炼方法能完成上述所有任务，某一种精炼方法只能完成其中一项或几项任务。由于各厂条件和冶炼钢种不同，一般是根据不同需要配备一两种炉外精炼设备。

0.2.6 Selection of Refining Equipment

0.2.6 精炼设备的选择问题

0.2.6.1 Select Refining Equipment According to Steel Composition and Quality

0.2.6.1 依据钢种成分与质量选择精炼设备

The purpose of molten steel treatment and refining function, is the basic factor to be considered when selecting equipment. If we want to reduce the gas inclusion in molten steel, we should consider that the refining method should have good degassing function. To solve the problem of temperature drop in the process of molten steel treatment, we should consider the refining method with heat compensation heating function.

钢水处理的目的和精炼功能，是选择设备时首先应考虑的基本因素。例如，若要降低钢水中的气体夹杂，则应考虑该精炼法是否具有良好的脱气功能；若要解决钢水处理过程中的温降问题，则应考虑是否采用具有热补偿——加热功能的精炼方法等。

0.2.6.2 Reference Steelmaking System Process Matching

0.2.6.2 参考炼钢系统流程匹配

The matching and connection of Primary refining furnace—Refining furnace—Continuous casting machine is a very important problem. The afteradded refining system should promote continuous casting and rolling mill, through temperature, flow, time rhythm, steel composition and purity.

初炼炉—精炼炉—连铸机的匹配与衔接是一个非常重要的问题。后加的炉外精炼系统应通过温度、流量、时间节奏、钢水成分和纯净度的衔接与匹配来促进连铸连浇，促进连铸机和轧机的衔接和匹配。

0.2.6.3 Focus on Operating Costs

0.2.6.3 关注运行成本

On the premise of ensuring safety and quality, the refining process should be simplified as much as possible. Although there are dozens of refining methods outside the furnace, there are not many common ones. According to different steel grades and processes, different refining methods are adopted. According to product quality, process and market requirements, different types of factories, different scales and products are equipped with different refining technologies.

在保证安全质量的前提下，尽量简化精炼环节。尽管炉外精炼方法多达几十种，但比较普遍的、常用的并不多，针对不同钢种和不同流程，采用不同的炉外精炼方法，根据产品质量、工艺和市场要求，对不同类型的工厂、规模和产品配备不同的精炼技术。

The large-scale joint enterprise producing plate and strip steel should be equipped with CAS-OB based compound refining station and RH vacuum treatment based refining station.

生产板带类钢材大型联合企业，应配备以 CAS-OB 为主的复合精炼站及以 RH 真空处理为主的精炼站。

The medium-sized iron and steel enterprises mainly producing rod and wire, should be equipped with LF furnace, wire feeding and other refining means.

对生产棒线材为主的中型钢铁企业，应配备 LF 炉、喂线等精炼手段。

For the special steel plants of electric furnace, EAF—AOD—VCR process is used to produce stainless steel; EAF—LF/VOD process is used to produce stainless steel and alloy structural steel; EAF—LF—RH—CC (continuous casting) or SKF—MR is used to produce bearing steel; EAF—LF/VD process is used to produce ultra pure structural steel.

电炉特殊钢厂生产不锈钢采用 EAF—AOD—VCR 工艺；生产不锈钢与合金结构钢采用 EAF—LF/VOD 工艺；生产轴承钢采用 EAF—LF—RH—CC（连铸）或 SKF—MR 工艺；生

产超纯结构钢采用 EAF—LF/VD 工艺。

The configurations of common steel grades are as follows（常见钢种的配置如下）：

(1) Alloy steel（合结钢）：EAF（或 BOF）—LF—CC

(2) Ball bearing（轴承钢）：EAF（EBT）（或 BOF）—LF—VD—CC

　　　　　　　　　　　　　EAF（EBT）（或 BOF）—LF—RH—CC

(3) Stainless steel（不锈钢）：EAF or BOF—VOD—(LF)—CC

　　　　　　　　　　　　　　EAF or BOF—AOD—(LF)—CC

　　　　　　　　　　　　　　EAF or BOF—AOD—VOD—(LF)—CC

0.2.7 Development Trend of Secondary Refining

0.2.7 炉外精炼技术的发展趋势

In production, all the devices of secondary refining are running online. In Japan, the radio of BOF with secondary refining was 70.8% in the 15 years ago, the radio of stainless steel was 94%, and all the DRI—EAF use secondary refining. It is economic and effective for getting different function with different divce of refining.

在实际生产中，炉外精炼设备全部在线运行。早在 15 年前，日本转炉钢的炉外精炼比就已达到 70.8%，特殊钢生产的二次精炼比高达 94%。新建电炉短流程钢厂采用炉外精炼。不在一个精炼炉中完成多项精炼任务，而是在不同精炼装置中分别完成，以便最经济、最有效地发挥不同精炼工艺的功能，实现炉外精炼的多功能化。

(1) The refining station is used for the connection of Converter—Continuous casting production with ladle argon blowing as the core, combined with one or more technologies such as wire feeding, powder spraying, chemical heating, fine adjustment of alloy composition, etc.

(1) 以钢包吹氩为核心，加上与喂线、喷粉、化学加热、合金成分微调等一种或多种技术相结合的精炼站，可用于转炉—连铸生产衔接。

(2) The refining station with vacuum treatment device as the core and combined with one or more of the above technologies, is mainly connected with Converter—Continuous casting production.

(2) 以真空处理装置为核心，并与上述技术中一种或几种技术复合的精炼站，主要与转炉—连铸生产衔接。

(3) Refining with ladle furnace (LF) as the core, combined with one or several technologies such as ladle argon blowing technology and vacuum treatment, is mainly used for the connection of electric furnace and continuous casting production.

(3) 以钢包炉 (LF) 为核心，与钢包吹氩技术及真空处理一种或几种技术相复合的精炼，主要用于电炉—连铸生产衔接。

(4) Stainless steel refining technology with AOD as the main body, including VOD and top bottom combined blowing of converter.

(4) 以 AOD 为主体，包括 VOD、转炉顶底复吹在内的不锈钢精炼技术。

Exercises

练习题

(1) What is the definition of out of furnace refining?

(1) 炉外精炼的定义是什么？

(2) What is the task of refining outside the furnace?

(2) 炉外精炼的任务是什么？

(3) What functions are generally required of the refining equipment outside the furnace?

(3) 一般要求炉外精炼设备具备哪些功能？

(4) What are the common means of refining out of furnace? How many types of refining equipment are there?

(4) 炉外精炼常用的手段有哪些，精炼设备通常可分为哪几类？

(5) Introduce the main development trend of refining technology outside the furnace.

(5) 简述当前炉外精炼技术的主要发展趋势。

Project 1　Refining Means and Process selection of Secondary Refining

项目 1　炉外精炼手段与工艺选择

The main means of refining outside the furnace include slag washing, stirring, vacuum, heating (temperature regulation) and blowing (wire feeding). Through the different combinations of these refining methods, the best thermodynamic and dynamic conditions are created for completing a certain refining task, and the refining equipment with different functions is composed, as shown in Table 1-1.

炉外精炼的主要手段包括渣洗、搅拌、真空、加热（调温）和喷吹（喂丝）。为完成某种精炼任务创造最佳热力学和动力学条件，通过这几种精炼手段的不同组合，构成功能不同的炉外精炼设备，其精炼手段、功能和代表工艺见表 1-1。

Table 1-1　Main means, functions and representative processes of external refining
表 1-1　炉外精炼主要手段、功能及代表工艺

Refining 精炼手段	Functions 功能	Refining means 代表工艺
Slag washing 渣洗	Desulfurization, deoxidation, and removal of impurities 脱 S，脱 O，去夹杂	Same furnace, different furnace and mixed VSR 同炉、异炉、混合 VSR
Vacuum 真空	Deaeration, desulfurization, and deoxidation 脱气，脱 C，脱 O	RH、DH
Mixing 搅拌	Purification, uniform composition and promotion of other metallurgical reactions 净化、均匀成分，促进其他冶金反应	Homogeneous composition 所有精炼方法
Heating temperature 加热	Adjust temperature and ensure refining time 调温，保证精炼时间	LF、SKF、CAS-OB、AOH
Injecting 喷吹	High efficiency addition of reactant 高效加入反应剂	—

The refining means of external refining are as follows:

以下分别介绍炉外精炼的主要手段：

(1) Slag washing: The synthetic slag prepared in advance (melted in a special slag smelting furnace) is poured into the ladle, and the impact of steel flow is borrowed to fully mix the molten steel and the synthetic slag, so as to complete the refining tasks such as deoxidization, desulfurization and removal of inclusions.

(1) 渣洗：将事先配好（在专门炼渣炉中熔炼）的合成渣倒入钢包内，借出钢时钢流的冲击作用，使钢液与合成渣充分混合，从而完成脱氧、脱硫和去除夹杂等精炼任务。

(2) Vacuum: The liquid steel is placed in a vacuum chamber, and the reaction moves towards the gas phase due to the vacuum action, so as to achieve the purpose of degassing, deoxidizing and decarbonizing. Vacuum is one of the most widely used means in refining.

(2) 真空：将钢液置于真空室内，由于真空作用使反应向生成气相方向移动，从而达到脱气、脱氧、脱碳等目的。真空是炉外精炼中广泛应用的一种手段。

(3) Agitation: the reaction interface by agitation is expanded, the transfer process of reaction substances is accelerated, and the reaction speed is improved. The stirring method includes air blowing and electromagnetic stirring.

(3) 搅拌：通过搅拌扩大反应界面，加速反应物质的传递过程，提高反应速度。其搅拌方法包括吹气搅拌和电磁搅拌。

(4) Heating: An important means to adjust the temperature of molten steel, so as to better connect steelmaking and continuous casting. The heating methods includes arc heating and chemical heating.

(4) 加热：调节钢液温度的一项重要手段，使炼钢与连铸更好地衔接。其加热方法包括电弧加热法和化学加热法。

(5) Blowing: A means of adding reactant into liquid metal with gas as carrier. The metallurgical function of the injection depends on the type of refining agent. It can complete the refining tasks such as decarburization, desulfurization, deoxidization, and alloying and controlling the shape of inclusions.

(5) 喷吹：用气体作载体将反应剂加入金属液内的一种手段。喷吹的冶金功能取决于精炼剂的种类，它能完成脱碳、脱硫、脱氧、合金化和控制夹杂物形态等精炼任务。

Various refining methods are basically composed of different combinations of these five refining methods (shown in Figure 1-1). One or more simple means can be used to form a unique refining technology. Different refining technologies meet different processing purposes and requirements. If a single secondary refining equipment can not meet the production requirements of steel varieties, two refining equipment duplex is required, such as LF—RH.

不同的精炼方法基本上由五种手段组成（见图1-1），一种技术运用一种或多种手段。不同的精炼技术满足不同的目的和要求，如果一种精炼设备不能满足不同钢种的生产要求，就需要两种精炼设备，如 LF—RH 双联。

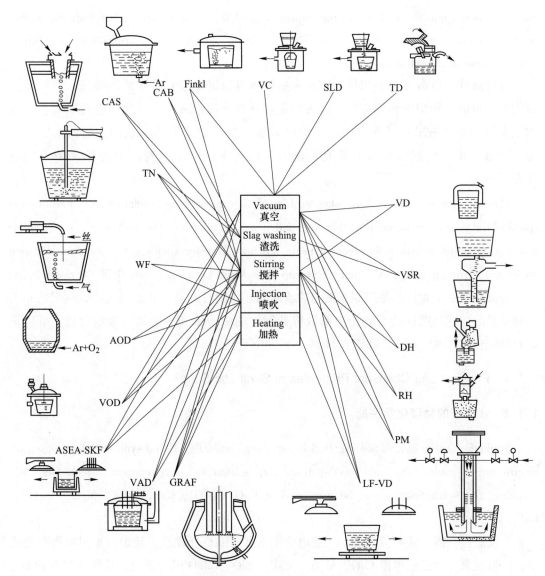

Figure 1-1 Main technologies of refining outside the furnace
图 1-1 炉外精炼主要技术

Task 1.1 Synthetic Slag Washing
任务 1.1 合成渣洗

The so-called synthetic slag washing is a kind of secondary refining method to further improve the quality of molten steel by washing the synthetic slag from the molten steel in the ladle. The most simple refining method is to obtain clean liquid steel and properly deoxidize, desulfurize and remove inclusions. The synthetic slag prepared in advance (melted in a special slag smelting furnace) is poured into the ladle, and the impact of steel flow is borrowed to fully mix the molten steel and the synthetic slag, so as to complete the refining tasks such as deoxidization, desulfurization and removal of inclusions.

合成渣洗是指由炼钢炉初炼的钢水在盛钢桶内通过钢液对合成渣的冲洗，进一步提高钢水质量的一种炉外精炼方法。该方法能够获得洁净钢液，并能适当进行脱氧、脱硫和去夹杂。将事先配好（在专门炼渣炉中熔炼）的合成渣倒入钢包内，借助出钢时钢流的冲击作用，使钢液与合成渣充分混合，从而完成脱氧、脱硫和去除夹杂等精炼任务。

The main purpose of synthetic slag washing is to control the deoxidation and desulfuration speed of the interaction between molten steel and slag, and to reduce the content of oxygen, sulfur and non-metallic inclusions in steel by using the mass transfer coefficient of material diffusion speed and stirring power. $w[O]$ and $w[S]$ can be reduced to 0.002% and 0.005% respectively.

合成渣洗的主要目的是控制钢液与炉渣间相互作用时的脱氧和脱硫速度，利用钢和渣中物质扩散速度及搅拌功率的传质系数降低钢中的氧、硫和非金属夹杂物含量，可以把 $w[O]$ 降至 0.002%，$w[S]$ 降至 0.005%。

1.1.1 Physical and Chemical Properties of Synthetic Slag

1.1.1 合成渣的物理化学性能

Synthetic slag includes solid slag and liquid slag. Generally, liquid synthetic slag is used for electric furnace molten steel and solid synthetic slag is used for converter molten steel. Generally speaking, the synthetic slag can be composed of CaO, Al_2O_3, CaF_2, SiO_2, MgO, Na_2CO_3, CaF_2, etc.

合成渣包括固态渣和液态渣，一般电炉钢水多用液态合成渣、转炉钢水多用固态合成渣。一般说来，合成渣可由 CaO、Al_2O_3、CaF_2、SiO_2、MgO 组合而成，其他还有 Na_2CO_3、CaF_2 等。

The synthetic slag mainly includes CaO-Al_2O_3 system, CaO-SiO_2-Al_2O_3 system, CaO-SiO_2-CaF_2 system, etc. At present, the commonly used synthetic slag system is mainly CaO-Al_2O_3 alkaline slag system. According to the CaO slagging effect and the chemical composition, it is roughly as follows: Lime slag with $w(CaO)=50\%\sim55\%$, high Alumina slag with $w(CaO)=40\%\sim45\%$, and refined slag with $w(CaO)=40\%\sim50\%$, as shown in Table 1-2.

合成渣主要有 CaO-Al_2O_3 系，CaO-SiO_2-Al_2O_3 系，CaO-SiO_2-CaF_2 系等。目前常用的合成渣系主要是 CaO-Al_2O_3 碱性渣系。依据 CaO 造渣作用，按照化学成分大致为：$w(CaO)=50\%\sim55\%$ 的石灰渣、$w(CaO)=40\%\sim45\%$ 的高氧化铝渣，$w(CaO)=45\%\sim50\%$ 精炼渣，见表 1-2。

Table 1-2 Main components of synthetic slag

表 1-2 合成渣主要成分

Slag type 渣类型	Components (mass fraction)/% 化学成分（质量分数）/%							
	CaO	MgO	SiO_2	Al_2O_3	FeO	Fe_2O_3	CaF_2	S
Lime clay slag	51.0	1.88	19.0	18.3	0.6	0.12	3.0	0.48
Lime alumina	48.94	4.0	6.5	37.83	0.74	—	—	0.63
Synthetic slag	46~50	—	3~12	22~26	—	—	—	—

In order to obtain the best refining effect, the synthetic slag is required to have the corresponding physical and chemical properties, and the composition of the slag is the decisive factor of the physical and chemical properties of the slag. The synthetic slag mainly includes $CaO-Al_2O_3$ system, $CaO-SiO_2-Al_2O_3$ system, $CaO-SiO_2-CaF_2$ system, etc. CaO content fluctuates (by mass) in the range of 45%~60%, MgO in the range of 6%~10%, Al_2O_3 in the range of 12%~16%, and SiO_2 in the range from 16% to 20%.

为了取得最佳的精炼效果，要求合成渣具备相应的物理化学性质，而炉渣的成分是炉渣物理化学性质的决定性因素。合成渣主要有 $CaO-Al_2O_3$ 系、$CaO-SiO_2-Al_2O_3$ 系和 $CaO-SiO_2-CaF_2$ 系等。其中 CaO 含量（质量分数）为 45%~60%，MgO 含量（质量分数）为 6%~10%，Al_2O_3 含量（质量分数）为 12%~16%，SiO_2 含量（质量分数）为 16%~20%。

CaO is a compound used for metallurgical reaction in synthetic slag, and other compounds are mostly added for adjusting composition and reducing melting point. The requirements for synthetic slag are as follows:

CaO 是合成渣中用于达到冶金反应目的的化合物，其他化合物多是为了调整成分，降低熔点而加入的。合成渣的要求有：

(1) It has high basicity and reducibility, low melting point, and good fluidity;

(1) 具有较高的碱度、高还原性、低熔点和良好的流动性；

(2) Suitable density, diffusion coefficient, surface tension and conductivity, etc.

(2) 具有合适的密度、扩散系数、表面张力和导电性等。

For the secondary refining of rod and wire products, the slag with the following components is recommended (by mass): $w(CaO) = 50\%~55\%$, $w(MgO) = 6\%~10\%$, $w(SiO_2) = 15\%~20\%$, $w(Al_2O_3) = 8\%~15\%$, and $w(CaF_2) = 5.0\%$. The total amount of SiO_2, Al_2O_3 and CaF_2 is controlled between 35% and 40% (by mass).

对于棒线材类产品炉外精炼，推荐采用下述成分的渣：$w(CaO) = 50\%~55\%$，$w(MgO) = 6\%~10\%$，$w(SiO_2) = 15\%~20\%$，$w(Al_2O_3) = 8\%~15\%$，$w(CaF_2) = 5.0\%$。其中，SiO_2、Al_2O_3、CaF_2 三组元的总量（质量分数）控制在 35%~40%之间。

1.1.1.1 Melting Point of Synthetic Slag

1.1.1.1 合成渣的熔点

When refining molten steel with synthetic slag in ladle, the melting point of slag should be lower than that of molten steel washed by slag. In the process of refining molten steel with synthetic slag in ladle, liquid slag is generally used, so the melting point of synthetic slag should be lower than that of molten steel washed by slag.

在钢包内用合成渣精炼钢水时，渣的熔点应当低于被渣洗钢液的熔点；在钢包内用合成渣精炼钢液时，由于一般用液态渣，因此合成渣的熔点应当低于被渣洗钢液的熔点。

The melting point of steel can be calculated approximately by

钢的熔点的计算公式为：

$$T_f = 1538 - \Sigma \Delta T_j [j\%] \tag{1-1}$$

Where T_f——the melting point of steel;

ΔT_j——the temperature change caused by 1% alloy element;

[j%]——the content of alloy elements.

式中 T_f——钢的熔点；

ΔT_j——1%合金元素引起温度变化；

[j%]——合金元素的含量。

The melting points of various synthetic slags can be determined according to the composition of synthetic slags and the corresponding phase diagrams.

各种合成渣的熔点可根据合成渣的成分，利用相应的相图来确定。

As shown in Tables 1-3 and 1-4:

由表 1-3 和表 1-4 可知：

Table 1-3   Melting point of CaO-Al$_2$O$_3$ slag system with different components

表 1-3   不同成分的 CaO-Al$_2$O$_3$ 渣系合成渣的熔点

Components (mass fraction)/% 化学成分（质量分数）/%				Melting point /℃ 熔点/℃
CaO	Al$_2$O$_3$	SiO$_2$	MgO	
46	47.7	—	6.3	1345
48.5	41.5	5	5	1295
49	39.5	6.5	5	1315
49.5	43.7	6.8	—	1335
50	50	—	—	1395
52	41.2	6.8	—	1335
56~57	43~44	—	—	1525~1535

Table 1-4   Melting point of CaO-SiO$_2$-Al$_2$O$_3$-MgO slag system with different components

表 1-4   不同成分的 CaO-SiO$_2$-Al$_2$O$_3$-MgO 渣系合成渣的熔点

Components (mass fraction)/% 化学成分（质量分数）/%						Melting point /℃ 熔点/℃
CaO	MgO	CaO+MgO	SiO$_2$	Al$_2$O$_3$	CaF$_2$	
58	10	68.0	20	5.0	7.0	1617
55.3	9.5	65.8	19.0	9.5	6.7	1540
52.7	9.1	61.8	18.2	13.7	6.4	1465
50.4	8.7	59.1	17.4	17.4	6.1	1448

(1) In CaO-Al$_2$O$_3$ slag system, when w(Al$_2$O$_3$) = 48% ~ 56% and w(CaO) = 52% ~ 44%, its melting point is the lowest (1450 ~ 1500℃). When a small amount of SiO$_2$ and MgO exist in the slag, its melting point will further decrease. The effect of SiO$_2$ content on the melting point of CaO-Al$_2$O$_3$ system is not as obvious as that of MgO.

(1) 在 CaO-Al$_2$O$_3$ 渣系中，当 w(Al$_2$O$_3$) = 48% ~ 56%，w(CaO) = 52% ~ 44%时，其熔点最低（1450 ~ 1500℃）。当这种渣存在少量 SiO$_2$ 和 MgO 时，其熔点还会进一步下降。SiO$_2$ 含量对 CaO-Al$_2$O$_3$ 系熔点的影响不如 MgO 明显。

(2) When 6% ~ 12% (by mass) of MgO is added to the CaO-Al$_2$O$_3$-SiO$_2$ ternary slag system, the melting point can be reduced to 1500℃ or even lower. Adding CaF$_2$, Na$_3$AlF$_6$, Na$_2$O, K$_2$O etc., can also reduce the melting point.

(2) 当 CaO-Al$_2$O$_3$-SiO$_2$ 三元渣系中加入 6% ~ 12%的 MgO（质量分数）时，就可以使其熔点降到 1500℃甚至更低。加入 CaF$_2$、Na$_3$AlF$_6$、Na$_2$O、K$_2$O 等也能降低熔点。

(3) CaO-SiO$_2$-Al$_2$O$_3$-MgO slag system has strong ability of deoxidization, desulfurization and absorption of inclusions. When the viscosity is fixed, the melting point of the slag increases with the increase of w(CaO+MgO) in the slag.

(3) CaO-SiO$_2$-Al$_2$O$_3$-MgO 渣系具有较强的脱氧、脱硫和吸附夹杂的能力。当黏度一定时，这种渣的熔点随渣中 w(CaO+MgO) 总量的增加而升高。

1.1.1.2 Fluidity of Synthetic Slag

1.1.1.2 合成渣的流动性

The synthetic slag used for slag washing needs better fluidity. Fluidity is one of the important factors that affect the emulsification degree of slag in molten steel. The temperature of molten steel is generally greater than 1600℃ at the steelmaking temperature. The temperature of the synthetic slag can be adjusted and controlled in the slag furnace according to the requirements. The viscosity of the synthetic slag is above 1600℃, and its viscosity is less than that of 0.2Pa·s.

用作渣洗的合成渣要求具有较好的流动性。流动性是影响渣在钢液中的乳化程度的重要因素之一。在炼钢温度下（混冲时钢液温度一般都大于 1600℃，合成渣的温度可按要求在炼渣炉中调整控制，一般也都在 1650℃以上），其黏度小于 0.2Pa·s。

At the same temperature and under the condition of mixed flushing, increasing the fluidity of the synthetic slag can reduce the average diameter of the emulsified slag drop, thus increasing the contact interface between slag and steel. The viscosity of CaO-Al$_2$O$_3$ slag with different components at the steel-making temperature is shown in Table 1-5. It has been found that the viscosity of the slag system is the lowest when the temperature is 1490 ~ 1650℃, w(CaO) = 54% ~ 56%, w(CaO)/w(Al$_2$O$_3$) = 1.2. Adding no more than 10% (by mass) of CaF$_2$ and MgO can also reduce the viscosity of slag. For most of the synthetic slag, its viscosity is less than 0.2Pa·s at the steelmaking temperature.

在相同的温度和混冲条件下,提高合成渣的流动性,可以减小乳化渣滴的平均直径,从而增大渣钢接触界面。在炼钢温度下,不同成分的 CaO-Al$_2$O$_3$ 渣的黏度如表 1-5 所示。研究表明,当温度为 1490~1650℃, $w(CaO) = 54\% \sim 56\%$, $w(CaO)/w(Al_2O_3) = 1.2$ 时,该渣系合成渣的黏度最小。加入不超过 10% 的 CaF$_2$ 和 MgO(质量分数)时,也能降低渣的黏度。对于大部分合成渣,在炼钢温度下,其黏度小于 0.2Pa·s。

Table 1-5　Viscosity of CaO-Al$_2$O$_3$ slag with different components
表 1-5　不同成分的 CaO-Al$_2$O$_3$ 渣的黏度

Components (mass fraction)/% 化学成分(质量分数)/%			Viscosity of slag at different temperatures (℃)/Pa·s 不同温度(℃)下渣的黏度/Pa·s					
SiO$_2$	Al$_2$O$_3$	CaO	1500	1550	1600	1650	1700	1750
—	40	60	—	—	—	0.11	0.08	0.07
—	50	50	0.57	0.35	0.23	0.16	0.12	0.11
—	54	46	0.60	0.40	0.27	0.20	0.15	0.12
10	30	60	—	0.22	0.13	0.10	0.08	0.07
10	40	50	0.50	0.33	0.23	0.17	0.15	0.12
10	50	40	—	0.52	0.34	0.23	0.17	0.14
20	30	50	—	—	0.24	0.18	0.14	0.12
20	40	40	—	0.63	0.40	0.27	0.20	0.15
30	30	40	0.92	0.61	0.44	0.38	0.24	0.19

1.1.1.3　Surface Tension

1.1.1.3　表面张力

In the process of slag washing, although the interface tension between Steel—Slag and the interface tension between slag and inclusion directly play an important role For example, the interface tension between steel and slag determines the diameter of the emulsified slag drop and the speed of the slag drop floating up, and the interface tension between slag and inclusion affects the ability of the slag drop to absorb and assimilate the non-metallic inclusions. But the size of the interface tension is directly related to the surface tension of each phase. As shown in Table 1-6, the surface tension is a relatively important parameter affecting the slag washing effect. The surface tension of slag is generally 550~580dyn/cm. When general steel grades are washed with CaO-Al$_2$O$_3$ synthetic slag, the surface tension is 700~1200 dyn/cm.

在渣洗过程中,虽然直接起作用的是钢—渣之间的界面张力和渣与夹杂之间的界面张力,如钢—渣间的界面张力决定乳化渣滴的直径和渣滴上浮的速度,而渣与夹杂间的界面张力的大小影响着悬浮于钢液中的渣滴吸附和同化非金属夹杂的能力。但是界面张力的大小是与每一相的表面张力直接相关。从表 1-6 可以看出,表面张力是影响渣洗效果的一个较为重要的参数,熔渣的表面张力计算为熔渣的表面张力一般为 550~580dyn/cm。当用 CaO-Al$_2$O$_3$ 系合成渣洗一般钢种时,其表面张力为 700~1200dyn/cm。

Table 1-6 Surface tension of common oxides in slag

表 1-6 渣中常见氧化物的表面张力

Oxide 氧化物	CaO	MgO	FeO	MnO	SiO$_2$	Al$_2$O$_3$	CaF$_2$
Surface tension/N·m^{-1} 表面张力/N·m^{-1}	0.52	0.53	0.59	0.59	0.40	0.72	0.405

1.1.1.4 Reducibility

1.1.1.4 还原性

The refining task that requires slag washing determines that the slag used for slag washing is high basicity (R>2) and $w(\text{FeO})$<0.4%~0.8%. The refining task completed by slag washing determines that the slag used for slag washing is reductive, and the content of FeO in the slag is very low, generally lower than 0.3% (by mass).

要求渣洗所用的熔渣都是高碱度（R>2），一般 $w(\text{FeO})$≤0.4%~0.8%。渣洗所用的都是还原性熔渣，渣中 FeO 含量都很低，一般低于 0.3%（质量分数）。

1.1.2 Slag Washing Process

1.1.2 渣洗工艺

1.1.2.1 Classification

1.1.2.1 分类

According to the different refining methods of synthetic slag, the slag washing process can be divided into different slag washing and the same slag washing (shown in Table 1-7).

根据合成渣炼制的方式不同，渣洗工艺可分为异炉渣洗和同炉渣洗（见表 1-7）。

Table 1-7 Classification of synthetic slag by refining method

表 1-7 合成渣按炼制方式分类

Classification 分类	Features 特点	Application 应用
Alternative furnace 异炉渣洗	Slag washing uses a special slag furnace to make slag. When tapping, the molten steel is rushed into the ladle with liquid slag in advance 用专用炼渣炉炼渣，出钢时钢液冲进事先盛有液渣的钢包内	(1) It is suitable for many kinds of steel in electric furnace (1) 适用于电炉许多钢种 (2) The process is complex and the scheduling is inconvenient, so a special slag smelting furnace is required to cooperate (2) 工艺复杂、调度不便，需一台专用

Continued Table 1-7

Classification 分类	Features 特点	Application 应用
Same furnace 同炉渣洗	The effect of slag washing liquid slag and steel liquid refining in the same furnace 液渣和钢液同一炉炼制	(1) Effect is not as good as that of different slag washing (1) 效果不如异炉渣洗 (2) For carbon steel and general low alloy steel (2) 用于碳钢和一般低合金钢

The so-called different slag washing is to set up a special slag smelting furnace (generally using electric arc furnace) to make the slag with a certain ratio into liquid slag with a certain temperature, composition and metallurgical properties. When tapping, the liquid steel is rushed into the ladle with such liquid slag in advance to realize slag washing.

异炉渣洗是指设置专用的炼渣炉（一般使用电弧炉），将配比一定的渣料炼制成具有一定温度、成分和冶金性质的液渣，出钢时钢液冲进事先盛有这种液渣的钢包内，实现渣洗。

The same slag washing means that the liquid slag and molten steel washed by slag are smelted in the same furnace, the liquid slag has the composition and property of synthetic slag, and then the task of slag washing molten steel is finally completed through tapping. For example, the steel slag mixed out during EAF smelting, which is called the same slag washing, and also uses the slag washing principle (applicable to the traditional 'Old Third Stage' smelting process of EAF, big mouth deep pit and slag during tapping mixed out).

同炉渣洗是指渣洗的液渣和钢液在同一座炉内炼制，并使液渣具有合成渣的成分与性质，然后通过出钢最终完成渣洗钢液的任务，例如，电弧炉冶炼时的钢渣混出（称同炉渣洗）也是利用了渣洗原理（适用于传统电弧炉"老三期"冶炼工艺，出钢时大口深坑，钢渣混出）。

In addition, according to the composition of the used synthetic slag, it can be divided into solid slag washing and premelted slag washing (shown in Table 1-8).

另外按照所用合成渣的成分可分为固体渣渣洗和预熔渣渣洗，见表1-8。

Table 1-8 Classification of synthetic slag by composition classification features application effect

表1-8 合成渣按成分分类

Classification 分类	Features 特点	Application 应用效果
Solid slag 固体渣渣洗	Slag washing solid synthetic slag is added into ladle before or during tapping 固体合成渣渣料在出钢前或出钢过程中加入钢包中	(1) Simplify process and operation (1) 简化工艺和操作 (2) Effect is not stable (2) 效果不稳定 (3) Temperature loss is large (3) 温度损失大

Continued Table 1-8

Classification 分类	Features 特点	Application 应用效果
Premelted slag 预熔渣 渣洗	Slag washing and premelting slag are added into ladle before or during tapping 预熔渣在出钢前或出钢过程中加入钢包中	The melting speed of premelting slag is faster than that of solid slag, and the effect is good 预熔渣熔化速度比固体渣快，效果好

The cost of solid slag is lower than that of premelted slag. Raw materials are easy to obtain and operation is simple. However, slag is not easy to melt completely, segregation is easy to occur, performance is unstable, and hydration is easy to affect refining effect, increase liquid steel [H] and cause quality problems. Table 1-9 shows the main components of solid slag.

固体渣成本比预熔渣较低，原材料容易获得，操作简单。但渣不易完全熔化，易产生偏析，性能不稳定，易水化，从而影响精炼效果，使钢液 [H] 增加，引起质量问题。固体渣的主要成分见表 1-9。

Table 1-9 Main components of solid slag
表 1-9 固体渣主要成分

Components (mass fraction)/% 化学成分（质量分数）/%							
Al	Al_2O_3	SiO_2	MgO	MnO+FeO	H_2O	P	S
20~22	45~55	8~12	5~7	≤1	0.5~0.6	0.12	0.10

The premelted slag has uniform composition, stable performance, not easy to absorb water during storage, uniform slag formation, fast speed, less dust, small environmental pollution, strong ability to absorb non-metallic inclusions in steel, and high comprehensive desulfurization rate. The main components are shown in Table 1-10.

预熔渣成分均匀，性能稳定，储存时不易吸水，成渣均匀，速度快，粉尘少，对环境污染小，有较强的吸附钢中非金属夹杂物的能力，综合脱硫率高。其主要成分见表 1-10。

Table 1-10 Typical components of premelted slag
表 1-10 预熔渣典型成分

Components (mass fraction)/% 化学成分（质量分数）/%				
CaO	Al_2O_3	SiO_2	MgO	Al
45~52	41~46	≤5	2~5	0.23

1.1.2.2 Washing Process of Synthetic Slag

1.1.2.2 合成渣洗过程

Washing process of synthetic slag: Before tapping, pour the prepared synthetic slag into the steel ladle and move it to the bottom of the steelmaking furnace. During tapping, the molten steel stream impacts the synthetic slag and stirs it fully to make the molten steel fully contact with the synthetic slag and make the molten steel get the slag washing. The steel flow has a certain height (the height of mixed flushing is generally 3~4m) and speed, and the molten steel will be cleaned soon, so the molten steel has a certain impact force, which can make the Steel—Slag fully agitated and contacted.

合成渣洗过程：出钢前将准备好的合成渣倒入盛钢桶内并移至炼钢炉下，在出钢过程中，钢液流冲击合成渣，充分搅拌，使钢液与合成渣充分接触，使钢液得到渣洗。钢流有一定的高度（混冲高度一般为3~4m）和速度，钢水很快出净，因此钢水有一定的冲击力，能使钢—渣充分搅拌接触。

1.1.2.3　Precautions

1.1.2.3　注意事项

In order to make the steel slag fully agitated and contacted, it is necessary to tap with large tap hole and deep pit. That is to say, the molten steel has a certain impact force, and the steel flow has a certain height and speed, so that the molten steel can be cleaned as soon as possible.

为了使钢渣充分搅拌接触，需大出钢口、深坑出钢，即钢水有一定的冲击力，钢流有一定的高度和速度，使钢水尽快出净。

The height of mixed flushing is generally 3~4m, and the tapping time is short, such as 35~55s for 10t furnace, 45~50s for 20t furnace, and the amount of synthetic slag is generally 5%~6% of the steel water. Before tapping the molten steel for slag washing, the slag shall be removed before tapping.

混冲高度一般为3~4m，出钢时间要短，如10t炉子35~55s，20t炉子45~50s，合成渣用量一般为钢水量的5%~6%。进行渣洗的钢液出钢前应扒除炉渣，然后再出钢。

1.1.3　Application Technology

1.1.3　应用技术

1.1.3.1　Slag Control Technology

1.1.3.1　顶渣控制技术

For quite a period of time, metallurgical workers thought that refining outside the furnace was a combination of vacuum, heating and stirring, which underestimated the role of top slag in ladle.

相当一段时间内，冶金工作者认为炉外精炼就是真空、加热及搅拌的组合，低估了钢包中顶渣的作用。

For different refining purposes, it should have the best top slag composition. For example, in order to deep deoxidize and desulfurize, the basicity of slag R should be 3~5 [$R = w(CaO)/$

$w(SiO_2)$], $w[\sum(FeO)]<0.5\%$. For low Aluminum killed steel, CaO saturated top slag and low aluminum [$w[Al]<0.005\%$] molten steel are used for stirring, so that the final oxygen activity is less than 0.0005%.

对于不同精炼目的,应有其最佳顶渣成分。例如,为了深度脱氧及脱硫,应该使渣碱度 R 达到 3~5 [$R=w(CaO)/w(SiO_2)$],$w[\sum(FeO)]<0.5\%$。而对于低铝镇静钢,则采用 CaO 饱和的顶渣与低铝 [$w[Al]\leq0.005\%$] 钢水进行搅拌,使最终的氧活度不大于 0.0005%。

1.1.3.2 Slag Retaining Technology

1.1.3.2 挡渣技术

The basic premise of refining slag is to do well the slag blocking operation during tapping and to reduce the oxide slag entering the ladle as much as possible. At present, there are many kinds of slag retaining technologies for industrial production: slag ball, floating plug for slag retaining, pneumatic blowing slag stopper, siphon out steel mouth to keep slag, and eccentric bottom tapping.

做好出钢时的挡渣操作,尽可能地减少钢水初炼炉的氧化渣进入钢包内是发挥精炼渣精炼作用的基本前提。目前已经出现了许多种用于工业生产的挡渣技术,包括:挡渣球,浮动塞挡渣,气动吹气挡渣塞,虹吸出钢口挡渣以及偏心炉底出钢。

1.1.4 Effect and Deficiency

1.1.4 作用效果及不足

1.1.4.1 Refining Effect of Slag Washing

1.1.4.1 渣洗的精炼作用

The commonly used synthetic slag washing process is to smelt the synthetic slag in a special slag smelting furnace. According to the required composition, temperature and slag amount, the synthetic slag is poured into the ladle first, and then hoisted to the tapping position of the electric arc furnace or converter. The synthetic slag is emulsified by the impact of the tapping steel flow. The slag and steel are fully mixed, and then the emulsified slag drops float up to the steel surface to complete the desulfurization, deoxidization and removal of non-metallic inclusions of the steel liquid. The refining task is to intensify deoxidization, desulphurization and dephosphorization, accelerate the removal of impurities in the steel, change the morphology of inclusions, prevent molten steel from inspiring, reduce the temperature loss of molten steel, and form the foaming slag to achieve the purpose of submerged arc heating.

常用的合成渣洗工艺流程是在专门的炼渣炉中冶炼合成渣,按要求的成分、温度、渣量先倒入钢包内,然后吊到电弧炉或转炉出钢位置,靠出钢钢流的冲击使合成渣乳化,渣和钢充分接触混合，将乳化的渣滴上浮至钢液面，完成钢液的脱硫、脱氧、去除非金属夹杂等精炼任务，从而实现强化脱氧、脱硫、脱磷，改变部分夹杂物形态，加快钢中杂质的排除，防止钢水吸气，减少钢水温度散失，形成泡沫性渣达到埋弧加热。

Emulsification and Floatation of Synthetic Slag
合成渣的乳化和上浮

Under the impact of the steel flow, the synthetic slag poured into the ladle is divided into small slag drops and dispersed in the molten steel. The smaller the particle size is, the larger the surface area contacting the molten steel is, and the stronger the slag washing effect is. The emulsified slag droplets are stirred by the steel flow, collided, merged and finally floated upward.

倒入钢包内的合成渣在钢流的冲击下，被分裂成细小的渣滴并弥散于钢液中，粒径越小，与钢液接触的表面积越大，渣洗作用越强。乳化的渣滴随钢流搅动，不断碰撞、合并，最后上浮。

Deoxidation of Synthetic Slag
合成渣脱氧

In the process of slag washing, with the decrease of molten steel temperature, the equilibrium of deoxidization moves to deoxidization direction, which may react with dissolved oxygen in steel and carry out precipitation deoxidization; and some oxygen diffuses into the slag to deoxidize the molten steel, such as:

在渣洗过程中，随着钢液温度的下降，脱氧反应的平衡向脱氧方向移动，有可能与钢中溶解的氧反应而进行沉淀脱氧；还有一部分氧通过扩散进入渣中，从而使钢液脱氧。比如：

(1) Silicon deoxidation. The reaction of silicon deoxidation is written as

(1) 硅脱氧。硅脱氧的反应式为

$$[Si]+2[O] = (SiO_2) \qquad (1-2)$$

Therefore, the following measures are mainly taken to improve the deoxidization ability of Si in practical engineering:

因此，在实际工程上为提高 Si 脱氧能力主要采取如下措施：

1) Increase of basicity（提高碱度）:

$$(CaO)+(SiO_2) = CaO \cdot SiO_2 \qquad (1-3)$$

SiO_2 is consumed by reaction (1-3).

1) 通过反应 (1-3)，消耗硅脱氧产物 SiO_2。

2) Compound Deoxidizer.

2) 复合脱氧剂脱氧。

(2) Aluminum deoxidation. The reaction of aluminum deoxidation is written as

(2) 铝脱氧。铝脱氧的反应式为

$$[Al]+3[O] = (Al_2O_3) \qquad (1-4)$$

In order to improve the deoxidization capacity of Al, the following measures are mainly taken in the factory practice:

为提高 Al 的脱氧能力，在实际生产中工厂主要采取如下措施：

1) Increase of basicity. The reaction of increase of basicity is written as

1) 提高碱度。提高碱度所发生的反应为

$$(CaO)+(Al_2O_3)=\!\!=\!\!=CaO \cdot Al_2O_3 \qquad (1-5)$$

2) Change the existing form of Al_2O_3, inclusion denaturation, (The change from solid to liquid inclusions is easy to float up).

2) 改变 Al_2O_3 的存在形态，即夹杂物变性（由固态变为液态的夹杂易于上浮）。

3) Enhanced stirring (soft blowing Ar), and the flotation separation effect of deoxidized products was enhanced by stirring.

3) 加强搅拌（软吹Ar），脱氧产物的上浮分离效果因搅拌而得到加强。

(3) Diffusion deoxidation. In practical engineering, the limiting link is diffusion. The measures to strengthen diffusion are as follows:

(3) 扩散脱氧。实际工程中，限制性环节在于扩散。加强扩散的措施包括：

1) The interface of steel slag is increased thousands of times by using synthetic slag emulsification.

1) 采用合成渣乳化，使钢渣界面成千倍增大。

2) Strong agitation.

2) 强烈搅拌。

Removal of Inclusions by Slag

夹杂物除渣

The removal of inclusions by slag washing is mainly manifested in two aspects:

夹杂物去除渣洗去夹杂的作用主要表现在两方面：

(1) In the molten steel, the original inclusion collides with the emulsion slag drop, is absorbed by the slag drop. After assimilation, if floats up to remove with the slag.

(1) 钢液中，原有的夹杂与乳化渣滴碰撞，被渣滴吸附，同化而随渣滴上浮去除。

(2) It promotes the discharge of the secondary reaction products, thus reducing the amount of inclusions in the finished steel.

(2) 促进了二次反应产物的排出，从而使成品钢中夹杂数量减少。

Desulfurization with Synthetic Slag

合成渣脱硫

In the process of slag washing, S in molten steel and CaO in synthetic slag react to form CaS and remove it.

在渣洗过程中，钢液中的S与合成渣中的CaO作用生成CaS，从而被去除。

1.1.4.2　Process Defects of Synthetic Slag Washing

1.1.4.2　合成渣洗的工艺缺陷

The synthetic slag washing process does not need large-scale equipment and has low investment. It is known as the simplest refining method. However, it has the following disadvantages and needs to be improved gradually in the future production:

合成渣洗工艺不需要大型设备，投资小，因此被誉为最简单的精炼手段。但是该工艺存在如下缺点：

(1) Unstable effect.

(1) 效果不稳定。

(2) It is impossible to remove the gas; and when the slag material is not dry, it may also cause suction, so as to increase [H].

(2) 不能去除气体；当渣料不干燥时还可能造成吸气，使[H]增加。

(3) The sulfide form cannot be controlled. In order to control the sulfide form, inject calcium line after slag washing.

(3) 不能控制硫化物形态。要想控制硫化物形态，则需要在渣洗后再注入钙线。

(4) High temperature drop.

(4) 温降大。

(5) Erosion of refractories.

(5) 耐火材料的侵蚀。

(6) The synthetic slag must be heated to liquid state for use. So a set of special equipment for melting synthetic slag is needed, generally using electric arc furnace for melting.

(6) 必须把合成渣加热到液态使用，这样就需要增加一套专门熔化合成渣的设备，一般使用电弧炉熔化。

Task 1.2　Mixing

任务1.2　搅拌

Stirring the molten steel in the reaction vessel is the most basic and important means of refining outside the furnace. Based on the physicochemical theory of metallurgical process, with the help of gas, electromagnetic induction and mechanical methods, the movement of molten steel and slag is realized. Stirring provides kinetic energy for the molten steel and promote the convection movement of the molten steel in the refining reactor. The reaction interface is enlarged by stirring, the transfer process of reactants is accelerated, and the reaction speed is increased. The stirring of molten steel can improve the dynamic conditions of metallurgical reaction, strengthen the mass and heat transfer of reaction system, accelerate the metallurgical reaction, homogenize the composition and temperature of molten steel, and facilitate the aggregation, growth and floatation of inclusions.

对反应容器中的钢液进行搅拌，是炉外精炼的最基本、最重要的手段。以冶金过程物理化学理论为基础，借助气体、电磁感应和机械方法，实现钢液和熔渣运动。搅拌给钢液提供动能，促使钢液在精炼反应器中对流运动。通过搅拌扩大反应界面，加速反应物质的传递过程，从而提高反应速度。钢液搅拌可改善冶金反应动力学条件，强化反应体系的传质和传热，加速冶金反应，均匀钢液成分和温度，从而有利于夹杂物聚合长大和上浮排除。

In general, stirring provides a certain amount of energy to the reaction system to promote the flow of the melt in the system. The heat and mass transfer process in the melt is accelerated by convection to achieve the effect of mixing. There are three main methods of stirring: gas stirring, electromagnetic stirring and mechanical stirring. The commonly used method is argon stirring, which can achieve the purpose of degassing (N, H), removing inclusions, deoxidizing and decarbonizing (shown in Figure 1-2).

通常系统向反应体系提供一定的能量,促使该系统内的熔体产生流动。通过对流加速熔体内传热和传质过程,从而达到混匀的效果。搅拌的方法主要有气体搅拌、电磁搅拌和机械搅拌三种方法。通常使用的方法是吹氩搅拌,该方法可以达到脱气(N、H),去除夹杂和脱氧脱碳的目的(见图1-2)。

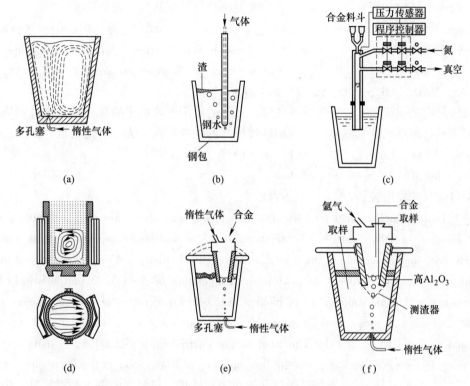

Figure 1-2 Secondary refining method with agitation
图 1-2 利用搅拌手段的炉外精炼方法
(a) Bottom blowing and stirring; (b) Embedded spray gun (SL); (c) Pulsating mixing (PM);
(d) Electromagnetic induction stirring; (e) Argon blowing on top (CAB);
(f) Sealing argon blowing (CAS)
(a) 包底吹气搅拌;(b) 埋入式喷枪(SL);(c) 脉动混合(PM);(d) 电磁感应搅拌;
(e) 封顶吹氩(CAB);(f) 密封吹氩(CAS)

1.2.1　Mixing Method

1.2.1　搅拌方法

1.2.1.1　Gas Mixing

1.2.1.1　气体搅拌

Argon is mainly used for gas agitation, so it is also called argon agitation. Argon gas can be blown into the molten steel by either top gun insertion method or bottom permeable brick method.

It has been proved by practice that the stirring effect can be fully exerted by blowing argon from the bottom through the permeable brick, and the utilization rate of argon is high. At present, most of the stirring methods of argon blowing are bottom blowing with permeable brick.

气体搅拌主要用氩气进行搅拌，故又称氩气搅拌。可以用顶枪插入法向钢液吹入氩气，也可以用底部透气砖法。实践证明，从底部通过透气砖吹入氩气，可充分发挥其搅拌作用，提高氩气的利用率。目前，大多数吹氩搅拌均采用透气砖底吹法。

Argon is the most common gas used to stir molten steel, and the use of nitrogen depends on the steel. Therefore, gas injection agitation is mainly various types of argon injection agitation. The refining methods used in this kind of agitation include ladle argon blowing, cab, CAS, VD, LF, Graf, VAD, VOD, AOD, SL, TN and so on.

氩气是用来搅拌钢水的最普通的气体，氮气的使用则取决于所炼钢种。因此喷吹气体搅拌主要是各种形式的吹氩搅拌。应用这类搅拌的炉外精炼方法有钢包吹氩、CAB、CAS、VD、LF、GRAF、VAD、VOD、AOD、SL、TN 等方法。

The roles of ladle argon are as follows:

钢包吹氩的作用包括：

（1）Uniform molten steel temperature: Due to the heat absorption of ladle lining and the heat dissipation of ladle surface, the temperature of molten steel around the ladle lining is relatively low, the temperature of the central area is relatively high, and the temperature of the upper and lower ladle is relatively low, while the temperature of the middle part relatively high, which results in the low temperature of molten steel before and after the pouring process of the tundish, and the high temperature in the middle part. The temperature of molten steel in ladle is stable and uniform, which is helpful to improve the internal quality of billet, to make the shell grow evenly in the mold, and to avoid the flow breaking of frozen steel in the open gate.

（1）均匀钢水温度：由于包衬吸热和钢包表面散热，包衬周围钢水温度较低，中心区域温度较高，钢包上、下部钢水温度较低，而中间温度较高，这种温度差异导致中间包浇注过程钢水温度前后期低，中期高。钢包吹氩搅拌促使钢包钢水温度稳定均匀，有利于提高铸坯内部质量，使结晶器内坯壳生长均匀，避免开浇水口冻钢断流。

（2）Uniform steel composition: In the process of argon blowing and stirring, fine-tuning of composition can be carried out according to the composition of molten steel provided by rapid analysis, so as to narrow the composition control range of steel and ensure the uniform performance of steel. The hydrogen and nitrogen content in the steel can be reduced and the oxygen content in the steel can be further reduced by using argon bubble gas to wash the steel water.

（2）均匀钢水成分：吹氩搅拌过程中可根据快速分析提供的钢水成分而进行成分微调，以使钢的成分控制范围更窄，从而确保钢材性能均匀。利用氩气泡气洗钢水能使钢中的氢、氮含量降低，并能使钢中的氧含量进一步下降。

（3）Promote inclusion floating, remove slag and inclusion, uniform temperature and composition, reduce segregation, and improve yield of deoxidizer and metal materials. The stirring molten steel promotes the non-metallic inclusions of steel to grow up. The floating argon bubble can absorb the gas in the steel, adhere the inclusions suspended in the molten steel and bring them to the surface of the molten steel to be absorbed by the slag layer. The production practice shows that the oxygen content of molten steel is obviously reduced after argon blowing and stirring, and the reduction range is related to the degree of deoxidation, which can be reduced by more than 20% generally. The amount of inclusions removed by argon blowing is related to the FeO content of the slag layer on the molten steel surface. The lower the FeO content in the slag, the more inclusions removed by argon blowing. In addition, the secondary oxidation of molten steel can be further avoided or reduced by using the protection of argon.

（3）促使夹杂物上浮，清除夹渣和夹杂，均匀温度和成分，减少偏析，从而提高脱氧剂和金属材料的收得率。搅动的钢水促使钢种非金属夹杂物碰撞长大，上浮的氩气泡能够吸收钢中的气体，同时黏附悬浮于钢水中的夹杂物并带至钢水表面被渣层所吸收。生产实践表明，吹氩搅拌后钢水氧含量明显降低，其降低幅度与脱氧程度有关，一般可降低20%以上。吹氩搅拌排除的夹杂物数量与钢水液面上覆盖渣层FeO含量有关，渣中的FeO含量越低，吹氩搅拌夹杂物的排除量越多。另外，利用氩气的保护作用，可进一步避免或减少钢液的二次氧化。

According to the circulation and flow of molten steel in the ladle, the movement of molten steel in the gas stirring ladle can be basically divided into four areas: A, B, C and D, as shown in Figure 1-3.

根据钢包内钢液的循环流动情况，气体搅拌钢包内钢液的运动基本上可以分为A、B、C、D四个区，如图1-3所示。

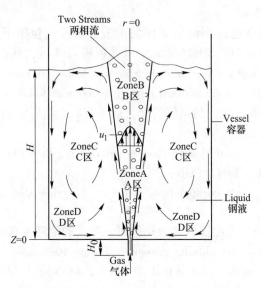

Figure 1-3　Movement of molten steel in ladle during gas stirring
图1-3　气体搅拌过程钢包内钢液的运动情况

The four areas (A, B, C and D) are be introduced respectively as follows:
以下分别对四个区（A区、B区、C区和D区）进行介绍：

（1）Zone A: It is a gas-liquid mixing area, which is the start-up area for the bubble to promote the liquid steel circulation. In this area, bubbles, molten steel and powder injected are

·37·

mixed well, with complex metallurgical reaction.

（1）A区：A区为气液混合区，是气泡推动钢液循环的启动区。在此区内气泡和钢液（若喷粉时还有粉料）充分混合，同时伴随着复杂的冶金反应。

（2）Zone B: After the gas-liquid stream in zone A rises to the top, the gas overflows and the liquid steel form a horizontal flow under the action of gravity, which disperse around. The molten steel and the scum on the top of the ladle form a two-phase liquid layer which is insoluble each other, and the slag layer and the molten steel layer slide at a certain relative speed.

（2）B区：在A区中气液流股上升至顶面以后，气体溢出而钢液在重力作用下形成水平流，向四周散开。呈放射形流散向四周的钢液与钢包中顶面的浮渣形成互不相溶的两相液层，渣层与钢液层之间以一定的相对速度滑动。

（3）Zone C: The liquid steel, horizontally and radially flows near the ladle wall and turns downward. The thickness of the zone is quite smaller than the ladle radius, because the liquid steel is dispersed around and constantly affected by the axial force of zone A in the process of downward flow.

（3）C区：水平径向流动的钢液在钢包壁附近，转向下方流动。由于钢液向四周散开，且在向下流动过程中又不断受到轴向A区的力的作用，所以该区的厚度与钢包半径相比较小。

（4）Zone D: The liquid steel returned to the lower part of the ladle along the ladle wall and the liquid steel in the middle and lower part of the ladle near zone A move from the periphery to the center under the action of the suction force in zone A. And they enter area A again to complete the circulation of liquid flow.

（4）D区：沿钢包壁返回到钢包下部的钢液，以及钢包中下部在A区附近的钢液，在A区抽引力的作用下由四周向中心运动。并再次进入A区，从而完成液流的循环。

According to the luck of gas stirring molten steel in ladle, the main factors influencing the argon blowing effects of ladle are summarized through practice as follows:

根据气体搅拌钢液在钢包内的运气情况，通过实践总结影响钢包吹氩效果的主要因素包括：

（1）The influence of argon blowing process parameters, such as:

（1）吹氩工艺参数的影响，包括：

1）The influence of argon consumption. When smelting low alloy steel, the consumption of argon is $0.2 \sim 0.4 m^3/t$.

1）氩气耗量的影响。在冶炼低合金钢种时，氩气的消耗量$0.2 \sim 0.4 m^3/t$。

2）The influence of argon blowing pressure. Taking 100t ladle as an example, when the argon blowing pressure is in the range of $0.2 \sim 0.5MPa$, the argon bubbles are distributed all over the ladle, and they are evenly distributed.

2）吹氩压力的影响。以100t钢包为例，当吹氩压力在$0.2 \sim 0.5MPa$范围内，氩气泡遍布整个钢包，呈均匀分布。

3）The influence of flow rate and argon blowing time. In different refining stages, the argon flow is different. In the argon blowing cleaning stage, the argon flow rate is $80 \sim 130L/min$; in the composition and slag mixing stage, the argon flow rate is $300 \sim 450L/min$; in the argon stirring stage, the argon flow rate is $450 \sim 900L/min$, and in the powder spraying stage, the argon flow rate is 900~1800L/min. Under normal conditions, the argon blowing time is 5~15 min, but the temperature of molten steel and the organization connection of the previous and next processes need to be considered.

3）流量和吹氩时间的影响。不同精炼阶段氩气流程不同。在吹氩清洗阶段，氩气流量为 80~130L/min；在调成分及化渣阶段，氩气流量为 300~450L/min；在精炼过程氩气搅拌阶段，氩气流量为 450~900L/min；在喷粉阶段，氩气流量为 900~1800L/min。在正常情况下，吹氩时间为 5~15min，但需要考虑钢液温度，以及上下工序的组织衔接。

（2）The influence of deoxidation degree, it is advisable to conduct argon blowing refining in ladle after good deoxidation treatment.

（2）脱氧程度的影响。钢包吹氩精炼要经过良好的脱氧处理后进行。

Combined with the current molten steel smelting process, there are two commonly used oxygen blowing methods: bottom blowing argon and top blowing argon (shown in Figure 1-4).

结合目前钢水冶炼工艺，常用的吹氧方式（底吹氩和顶吹氩见图 1-4）。

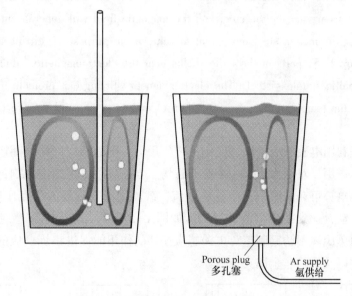

Figure 1-4　Top and bottom argon blowing
图 1-4　顶吹氩与底吹氩

The top blowing argon is to immerse the molten steel from the upper part of the ladle through the argon gun for argon blowing and stirring. It is required to set up a fixed argon blowing station. But the stirring effect of top blowing is not as good as bottom blowing.

顶吹氩是通过吹氩枪从钢包上部浸入钢水进行吹氩搅拌，要求设立固定吹氩站，但顶吹氩搅拌效果不如底吹氩好。

Most of the bottom blown argon is blown into the ladle through the permeable brick installed at a certain position at the bottom of the ladle. The advantages include: uniform molten steel temperature is uniformed, composition and removal of inclusions effect is good, the equipment is simple, and the operation is easy. Ladle bottom blowing argon stirring can be combined with other technologies to form a new refining methods outside the furnace. The disadvantage is that the permeable brick is easy to block sometimes, which is not synchronous with the ladle life.

底吹氩大多数是通过安装在钢包底部一定位置的透气砖吹入氩气。其优点包括：钢水温度均匀，成分和去除夹杂物效果好，设备简单，操作灵便。钢包底吹氩搅拌还可与其他技术配套组成新的炉外精炼方式。缺点是透气砖有时易堵塞，与钢包寿命不同步。

1.2.1.2　Electromagnetic Induction Stirring

1.2.1.2　电磁感应搅拌

Using the principle of electromagnetic induction to make molten steel move is called electromagnetic stirring. The magnetic field generated by the electromagnetic induction stirring coil can produce stirring effect in the molten steel. And an alternating magnetic field is applied to the molten steel. When the magnetic field cuts the molten steel at a certain speed, it will generate an induction potential, which can generate an induction current in the molten steel. And the interaction between the current carrying molten steel and the magnetic field will generate an electromagnetic force, thus driving the molten steel movement to achieve the purpose of stirring the molten steel. As shown in Figure 1-5, parts of the ladle shelks near the electromagnetic induction stirring coil are made of austenitic stainless steel. The electromagnetic stirring can promote the refining reaction, homogenize the temperature and composition of molten steel, and remove the non-metallic inclusions.

电磁搅拌是利用电磁感应的原理使钢液产生运动。电磁感应搅拌线圈产生的磁场可在钢水中产生搅拌作用，对钢水施加一个交变磁场。当磁场以一定速度切割钢液时，会产生感应电势，这个电势可在钢液中产生感应电流，载流钢液与磁场的相互作用产生电磁力，从而驱动钢液运动，达到搅拌钢液的目的。如图 1-5 所示，靠近电磁感应搅拌线圈的部分，钢包壳应由奥氏体不锈钢制造。电磁感应搅拌的作用能促进精炼反应的进行，均匀钢液温度和成分，以及去除非金属夹杂。

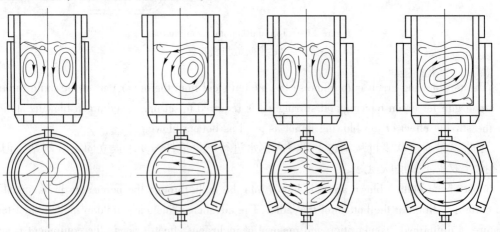

Figure 1-5　Effect of electromagnetic induction stirring under different forms of magnetic field operation

图 1-5　不同形式的磁场作业下电磁感应的搅拌效果

Gas agitation and electromagnetic induction agitation have their own characteristics, which should be reasonably selected according to the design capacity of iron and steel enterprises, and the level of energy consumption. However, from the perspective of investment, electromagnetic induction agitation needs to increase electromagnetic induction equipment, which has a large investment. Therefore, gas agitation (especially argon agitation), is widely used in secondary refining at present. The characteristics of the two processes are shown in Table 1-11.

气体搅拌与电磁感应搅拌各具有特点，应根据钢铁企业设计能力，能源消耗水平合理选择。但从投资看，由于电磁感应搅拌需增加电磁感应设备，投资较大，因而气体搅拌（特别是氩气搅拌）在目前炉外精炼使用较为普遍。气体搅拌与电磁感应搅拌工艺对比见表1-11。

Table 1-11　Comparison of gas stirring and electromagnetic induction stirring processes
表1-11　气体搅拌与电磁感应搅拌工艺对比

Project 项目	Blowing 吹气搅拌	Electromagnetic 电磁感应搅拌
Mixing capacity and regulation performance 搅拌能力及调节性能	The stirring capacity is not limited by the treatment capacity, and the blowing intensity is adjusted by the blowing capacity 搅拌能力不受处理容量限制吹气影响。通过吹气量可调节吹气强度	It depends on the power of the agitator and the working current. It can be adjusted by changing the current 取决于搅拌器功率大小工作电流，通过改变电流调节
Mixing uniformity 搅拌均匀性	Circulation is easy to form and nonuniform 容易形成环流，且不均匀性	It's better than blowing and stirring 比吹气搅拌均匀一些
Requirements for ladle 对钢包的要求	More consideration is given to the service life of refractories, which are generally designed as cone barrel type 更多考虑耐火材料的寿命，一般设计成锥桶型	It is required that the bath depth is large, and the furnace lining is as thin as possible and made of non magnetic steel 要求熔池深度较大，炉衬尽可能薄，无磁钢制成
Effect on refining reaction 对精炼反应的影响	The tendency of carburization is obvious and degassing is better than electromagnetic stirring to accelerate the temperature drop of molten steel 增碳倾向明显、脱气方面优于电磁搅拌加速钢液温降	Removal of inclusions is better than air blowing and stirring, which is not easy to produce carburization and has small temperature drop 去除夹杂优于吹气搅拌，不容易产生增碳，温降小
Investment 投资	Small investment 投资小	Large investment and high maintenance cost 投资大，维护费用高

1.2.1.3　Circulation Mixing

1.2.1.3　循环搅拌

In addition to gas agitation and electromagnetic induction agitation, it is often used in vacuum refining to achieve refining effect. The typical circulation stirring is that the liquid steel is stirred in the process of circulation flow, such as RH and DH, also known as suction and discharge stirring.

在真空精炼手段中，除了运用气体搅拌和电磁感应搅拌外，也常运用循环搅拌实现精炼效果。典型的循环搅拌是钢液在循环流动的过程中实现搅拌，如 RH 与 DH 的搅拌方式，也称作吸吐搅拌。

In RH refining, the stirring in the ladle is caused by the liquid steel injected into the ladle in the vacuum chamber (called bubble pumping phenomenon). The liquid lifted by the bubble produced by gas injection is called bubble pumping phenomenon, as shown in Figure 1-6.

在 RH 精炼中，钢包内的搅拌是由真空室内钢液注流进入钢包中引起的（称作气泡泵起现象）。气泡泵起现象是指用喷吹气体所产生的气泡来提升液体的现象，如图 1-6 所示。

At present, bubble pumping has been widely used in chemical industry, thermal power, metallurgy and so on. The commonly used ladle argon stirring is actually a modified 'Bubble Pump', which creates a lifting zone of low density gas-liquid mixture above the nozzle, and promotes the circulation of molten steel in the ladle. In the vacuum circulation degassing method (RH method), the circulation process of liquid steel also has the function principle of 'Bubble Pump' (shown in Figure 1-7).

目前，气泡泵起现象已广泛应用于化工、热能动力、冶金等领域。常用的钢包吹氩搅拌实际上是变形的"气泡泵"，它在喷口上方造成了一个低密度的气液混合物的提升区，它推动了钢包中钢液的循环流动。真空循环脱气法（RH 法）中，钢液的循环过程也具有"气泡泵"的作用原理（见图 1-7）。

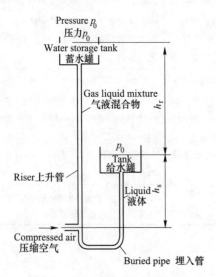

Figure 1-6 Schematic diagram of 'Bubble Pump'

图 1-6 "气泡泵"原理图

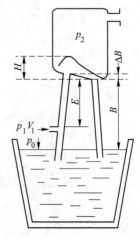

Figure 1-7 Circulation schematic diagram of RH 'Air Bag Pump'

图 1-7 RH "气包泵"循环原理图

1.2.2 Mixing and Homogenization

1.2.2 搅拌与混匀作用

It is considered that the stirring of the liquid steel is the result of the work done by the external force, the greater the energy input into the liquid steel to cause the stirring of the liquid steel, the more intense the stirring of the liquid steel will be. The stirring effects of liquid steel are measured by using the mixing time. The time needed to detect the uniform distribution of the testing agent in the stirred melt by adding tracer is called mixing time. The more intense the melt is stirred, the shorter the mixing time.

考虑到钢液的搅拌是由于外力做功的结果，所以在单位时间内，输入钢液内引起钢液搅拌的能量越大，钢液的搅拌越剧烈。通常利用混用时间去衡量钢液的搅拌效果。在被搅拌的熔体中加入示踪剂，检测出测试剂在熔体中均匀分布所需的时间称作混匀时间。熔体被搅拌得越剧烈，混匀时间就越短。

Of course, the fundamental mixing means that the composition or temperature is the same everywhere in the refining equipment, but this is almost impossible. But it can be imagined that, without any external interference, the more violent the melt is stirred, the shorter the mixing time.

当然，混匀理论上是指成分或温度在精炼设备内处处相同，但这几乎是做不到的。但可以设想，在没有任何外界干扰的情况下，熔体被搅拌得越剧烈，混匀时间就越短。

The mixing time will have a certain relationship with the metallurgical reaction rate, because most of the limiting links of metallurgical reaction rate are mass transfer. If the specific stirring power describing the stirring degree can be quantitatively related to the mixing time, the relationship between stirring and metallurgical reaction can be more clearly analyzed. The results obtained by different researchers are quite different, because the mixing of liquid steel is not only affected by the stirring power, but also by the diameter of the molten pool, the number of ventilating elements and other factors.

由于大多数冶金反应速率的限制性环节都是传质，所以混匀时间与冶金反应的速率有一定的联系。如果把描述搅拌程度的比搅拌功率与混匀时间定量联系起来，那么就可以较明确地分析出搅拌与冶金反应之间的关系。不同研究人员得到的研究结果之间有很大差别，这主要是因为钢液的混匀除了受搅拌功率的影响之外，还受熔池直径、透气元件个数等因素的影响。

Task 1.3　Heating

任务 1.3　加热

When the molten steel is refined outside the furnace, the temperature drops because of heat loss. If the furnace refining methods have the function of heating and warming up, they can avoid high temperature tapping, ensure the normal pouring of molten steel, and increase the flexibility of the furnace refining process and the amount of refining agent. And the final temperature and processing time of molten steel treatment can be freely selected to obtain the best refining effect. Therefore, many refining processes use heating as an important means to ensure the quality of molten steel, such as DH, SKF, LF, LFV, VAD, CAS-OB, etc.

钢液在进行炉外精炼时，由于有热量损失，造成温度下降。若炉外精炼方法具有加热升温功能，可避免高温出钢和保证钢液正常浇注，增加炉外精炼工艺的灵活性和精炼剂用量，钢液处理最终温度和处理时间均可自由选择，从而获得最佳的精炼效果。因此很多炉外精炼工艺都使用加热作为保证钢液质量的重要手段，如 DH、SKF、LF、LFV、VAD、CAS-OB 等。

There are two ways to reduce the temperature of molten steel in the refining process of the external refining device without heating means. One is to improve the tapping temperature, the other is to shorten the refining time outside the furnace. But it is difficult to ensure the quality of refining outside the furnace and the steel-making production organization. In addition, it needs the capacity of the ladle, the situation of slag coverage on the steel surface, the type and quantity of added materials, the method and strength of mixing, the structure of the ladle (the thermal conductivity of the ladle wall, whether the ladle has a cover) and the temperature before baking and other factors. So as an important means of secondary refining, the necessities of heating are as follows:

无加热手段的炉外精炼装置，精炼过程中钢液的降温常用以下两种办法来解决。一是提高出钢温度，另一种是缩短炉外精炼时间，但很难保证炉外精炼的质量以及炼钢生产组织。另外还需要考虑钢包的容量；钢液面上熔渣覆盖的情况；添加材料的种类和数量，搅拌的方法和强度；以及钢包的结构（包壁的导热性，钢包是否有盖）和使用前的烘烤温度等因素。因此加热作为炉外精炼的重要手段，其必要性如下：

(1) The temperature of molten steel drops from primary refining furnace to refining furnace.

(1) 钢液从初炼炉到精炼炉过程钢液温降。

(2) Heat is needed to melt slagging and alloy materials.

(2) 熔化造渣材料和合金材料需要热量。

(3) The temperature in vacuum degassing drops and argon stirring absorbs heat.

(3) 真空脱气时的温降和吹氩搅拌时氩气吸热。

(4) It is necessary to ensure the sufficient refining time and molten steel temperature.

(4) 需要保证足够充裕的精炼时间和钢液温度。

(5) It is necessary to ensure the proper pouring temperature of molten steel.

(5) 需要保证钢液具有合适的浇注温度。

1.3.1 Heating Method

1.3.1 加热方法

The commonly used heating methods include fuel combustion heating, electric heating and chemical heating. Fuel combustion heating is the use of fossil fuels (such as coal gas, natural gas, heavy oil, etc.), combustion heating as a heat source. Electric heating mainly includes arc heating and induction heating. The graphite electrode is used for arc heating. After being electrified, the arc is generated between the electrode and the molten steel. The molten steel is heated by the high temperature of the arc. In addition, the resistance is used for heating. Chemical heating is to use the chemical heat produced by exothermic reaction to heat the molten steel. It is a very effective heating method popular in recent years.

常用的加热方法有燃料燃烧加热、电加热和化学加热。燃料燃烧加热是利用矿物燃料（如较常用的是煤气、天然气、重油等），燃烧发热作为热源。电加热主要包括电弧加热和感应加热。电弧加热采用石墨电极，通电后，在电极与钢液间产生电弧，依靠电弧的高温加热钢液。此外还采用过电阻加热方法。化学加热是利用放热反应产生的化学热来加热钢液，是近年来流行起来的实效性很强的一种加热方法。

1.3.1.1　Fuel Combustion Heating

1.3.1.1　燃料燃烧加热

Fuel combustion and heating is the use of fossil fuels. The more commonly used fossil fuels are gas, natural gas, heavy oil, etc., which should be reasonably selected according to the local resource situation. The use of fuel combustion heating equipment is easy to use with the existing equipment in the smelting workshop, and it is easy to be quoted and mastered with low investment and mature technology, and the operation cost is also low. Figures 1-8 and 1-9 show the processes of using gas and solid fuel heating respectively. However, the fuel combustion heating methods are convenient to implement, but the disadvantages are obvious:

燃料燃烧加热是利用矿物燃料加热，较常用的矿物燃料有煤气、天然气、重油等。在选择燃料时应根据当地的资源情况合理选择。燃料燃烧加热的设备简单，很容易与冶炼车间现有设备配套使用，同时，燃料燃烧加热投资省，技术成熟，容易被引用和掌握，运行费用也较低。使用气体及固体燃料加热的工艺分别如图1-8和图1-9所示。燃料燃烧加热，实施起来较为方便，但也存在不足之处：

(1) When the liquid steel is heated, the oxygen potential of the liquid steel and the refining slag covered on the liquid steel will inevitably increase, which is not conducive to the desulfurization, deoxidization and other refining reactions. Because the combustion flame is oxidable, and the liquid steel is always expected to be in the reducing atmosphere during the refining outside the furnace.

(1) 由于燃烧的火焰是氧化性的，而炉外精炼时总是希望钢液处在还原性气氛下，这样在钢液加热时，必然会使钢液和覆盖在钢液面上的精炼渣的氧势提高，不利于脱硫、脱氧以及其他精炼反应的进行。

(2) When preheating a vacuum chamber or ladle furnace with an oxidizing flame, the refractory lining will be under the repeated and alternating action of oxidation and reduction, thus reducing the service life of the lining.

(2) 用氧化性火焰预热真空室或钢包炉时，会使其内衬耐火材料处于氧化、还原的反

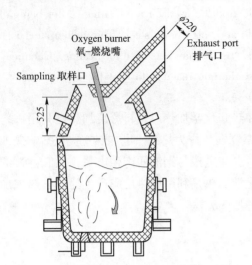 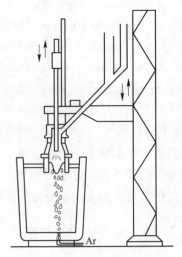

Figure 1-8　Gas fuel heating　　　　Figure 1-9　Solid fuel heating
图 1-8　气体燃料加热　　　　　　　图 1-9　固体燃料加热

复交替作用下，从而使内衬的寿命降低。

(3) It is inevitable that some residual steel will be stuck on the inner lining of vacuum chamber or ladle furnace. When the oxidation flame is used for preheating, the surface of these residual steel will be oxidized. In the next refining, these oxidized residual steel will become one of the sources of secondary oxidation oxygen of the refined steel liquid.

(3) 真空室或钢包炉内衬上不可避免会粘上一些残钢。当使用氧化性火焰预热时，这些残钢的表面会被氧化，而在下一炉精炼时，这些被氧化的残钢就被精炼钢液二次氧化。

(4) The partial pressure of water vapor in the flame will be higher than that under normal conditions. Especially when the fuel containing hydrocarbon is burned, which will increase the possibility of hydrogen generation from the refined steel.

(4) 火焰中的水蒸气分压将会高于正常情况下的水蒸气分压，特别是燃烧含有碳氢化合物的燃料时，这样将增大被精炼钢液增氢的可能性。

(5) A large amount of flue gas (combustion products) after fuel combustion makes these heating methods not easy to use with other refining methods (especially vacuum).

(5) 燃料燃烧后会产生大量烟气（燃烧产物），使得这种加热方法不利于与其他精炼手段（特别是真空）配合使用。

1.3.1.2　Electric Heating

1.3.1.2　电加热

Electric energy, as a convenient resource, is also used as a heating method for refining outside the furnace. The electric heating modes are mainly based on the selection of various heating means, which include:

电能可在炉外精炼的加热方法过程中使用。电加热方式主要根据选用各种不同加热手段进行加热，主要包括：

(1) Arc heating: The principle of arc heating is similar to that of electric arc furnace. The graphite electrode is used. After being electrified, the arc is generated between the electrode and the molten steel, and the molten steel is heated by the high temperature of the arc. Due to the high temperature of the arc, it is necessary to control the arc length and make foaming slag for submerged arc heating in the heating process, so as to prevent the high temperature erosion of the refractory caused by the arc. For the refining methods of arc heating under atmospheric pressure (such as LF, VAD, ASEA-SKF, etc.), the heating time should be shortened as much as possible to reduce the time of secondary gas absorption of molten steel. The schematic diagram of arc heating is shown in Figure 1-10.

(1) 电弧加热：电弧加热原理与电弧炉相似，都采用石墨电极。通电后，在电极与钢液间产生电弧，依靠电弧的高温加热钢液。由于电弧温度高，在加热过程中，需控制电弧长度，造好发泡渣进行埋弧加热，从而防止电弧对耐火材料产生高温侵蚀。常压下电弧加热的精炼方法（如 LF、VAD、ASEA-SKF 等），加热时间应尽量缩短，从而减少钢液二次吸气的时间。电弧加热示意图如图 1-10 所示。

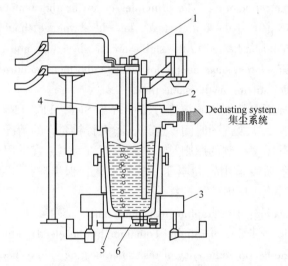

Figure 1-10　Schematic diagram of arc heating
图 1-10　电弧加热示意图
1—Motor; 2—Spray gun; 3—Ladle car; 4—Ladle cover; 5—Argon gas pipe; 6—Sliding nozzle
1—电机；2—喷枪；3—钢包车；4—钢包盖；5—氩气管；6—滑动水口

There are three electrode holes (such as alloy adding holes, exhaust gas discharging holes, and sampling and temperature measuring holes) on the ladle cover. If necessary, spray gun holes are also required. Arc heating is generally powered by a special three-phase transformer. The whole set of power supply system, control system, detection and protection system, as well as the way of arcing are the same as the general electric arc furnace, the difference is that the unit capacity of the transformer (average transformer capacity per ton of refined steel liquid) is smaller, the secondary voltage classification is more, the electrode diameter is smaller, the current density is large, and the quality of the electrode is required to be high.

在钢包盖上有 3 个电极孔（添加合金孔，废气排放孔，以及取样和测温孔），如果有必要，还需装设喷枪孔。电弧加热一般由专用的三相变压器供电。整套供电系统，控制系统，检测和保护系统，以及燃弧的方式都相同于一般的电弧炉，其不同之处在于配用的变压器单位容量（平均每吨被精炼钢液的变压器容量）较小，二次电压分级较多，电极直径较细，电流密度大，以及对电极的质量要求高等方面。

It is difficult for the arc heating methods to solve the problems such as the high performance of the electrode, the short distance between the arc and the lining of the ladle furnace, the short service life of the lining, and the promotion of liquid steel suction when the arc is heated under normal pressure.

电弧加热法难以解决一些问题，比如：电弧加热对电极的性能要求太高，电弧距钢包炉内衬的距离太近，包衬寿命短，常压下电弧加热时促进钢液吸气等。

Therefore, for the refining methods of electric arc heating under atmospheric pressure (such as SKF, LF, LFV, etc.), the heating time should be shortened as much as possible to reduce the time of secondary gas absorption of liquid steel; and the maximum heating rate of the refining should be allowed by the refractories. The principles of arc heating temperature control include: slag forming is the main part in the initial stage, low-level voltage should be adopted, and the middle level current should be heated to the stable arc; high voltage and current are used in the heating stage; and the insulation stage adopts low voltage and medium current. Cut off the cooling clamp and blow argon for stirring at the same time.

因此，在使用常压下电弧加热的精炼方法中（如 SKF、LF、LFV 等），加热时间应尽量缩短，从而减少钢液二次吸气的时间；应该在耐火材料允许的情况下，使精炼具有最大的升温速率。电弧加热温度控制的原则是：初期以造渣为主，宜采用低级电压，中档电流加热至电弧稳定；升温阶段采用较高电压，较大电流；保温阶段采用低级电压，中小电流。降温夹断时必须停电，同时吹氩搅拌。

(2) Resistance heating: The graphite resistance bar is used as the heating element to heat the steel liquid or the inner lining of refining container by the electric current and the resistance heat of graphite bar. The heating efficiency of resistance heating is low, because the heating method relies on radiation heat transfer. For example, the DH method can slow down or prevent the cooling of molten steel in the refining process after using resistance heating. It is very difficult to obtain a practical temperature raising rate through this heating method.

(2) 电阻加热：电阻加热利用石墨电阻棒作为发热元件，通以电流，靠石墨棒的电阻热来加热钢液或精炼容器的内衬。因为这种加热方法是靠辐射传热所以电阻加热的加热效率较低。例如，DH 法使用电阻加热后，可减缓或阻止精炼过程中钢液的降温。通过电阻加热方法获得有实用价值的提温速率是极为困难的。

1.3.1.3 Chemical Heating

1.3.1.3 化学加热

Chemical heating uses the chemical heat produced by exothermic reaction to heat the molten steel. The commonly used methods are silicothermic, aluminothermic and co-secondary combustion method. Chemical heating requires blowing oxygen to react with silicon, aluminum and CO to generate heat.

化学加热是利用放热反应产生的化学热来加热钢液。常用的方法有硅热法、铝热法和 CO 二次燃烧法。化学加热需吹入氧气，与硅、铝、CO 反应，从而产生热量。

Oxygen is blown into the steel by an oxygen gun, which reacts with the heating agent in the steel to produce chemical heat. The combustion heat is transmitted to the molten steel by radiation, conduction and convection, and then to the deep part of the molten steel by means of argon stirring. Generally, the top blowing oxygen gun is used in the chemical heating method, and the common blowing oxygen gun is of consumption type, which is composed of double-layer stainless steel pipe. The outer lining is made of high alumina refractory $[w(Al_2O_3) \geqslant 90\%]$, and the casing clearance is generally 2~3mm. The outer tube is cooled by argon, which accounts for about 10% of oxygen (by mass). The burning rate of oxygen lance is about 50 mm/time, and its life is 20~30 times.

利用氧枪吹入氧气，与加入钢中的发热剂发生氧化反应，从而产生化学热。燃烧热通过辐射、传导、对流传给钢水，借助氩气搅拌将热传向钢水深部。一般在化学加热法中多采用顶吹氧枪，常见的吹氧枪为消耗型，用双层不锈钢管组成。外衬高铝耐火材料 $[w(Al_2O_3) \geqslant 90\%]$，套管间隙一般为 2~3mm。外管通以氩气冷却，氩气量大约占氧量的 10% 左右（质量分数）。氧枪的烧损速度大约为 50mm/次，寿命为 20~30 次。

There are two kinds of heating agents used in chemical heating process. One is metal heating agent, such as aluminum, silicon, manganese, etc. The other is alloy heating agent, such as Si-Fe, Si-Al, Si-Ba-Ca, Si-Ca, etc. Aluminum and silicon are the preferred heating agents.

对于化学加热过程使用的发热剂主要有两大类，一类是金属发热剂，如铝、硅、锰等；另一类是合金发热剂，如 Si-Fe、Si-Al、Si-Ba-Ca、Si-Ca 等。铝和硅是首选的发热剂。

The Aluminum—Oxygen heating method (AOH) of molten steel is a kind of chemical heat method. It uses the lance to blow oxygen to oxidize the dissolved aluminum in the steel and release a large amount of chemical heat, which makes the molten steel rise rapidly. The method has many advantages: It rarely produces smoke, because the lance is immersed in the molten steel during oxygen blowing; It can accurately predict the temperature rise result, because oxygen is all in direct contact with the molten steel; and it has no effect on the ladle life, and also can obtain the molten steel with high cleanliness. CAS-OB and RH-OB also have similar function of AOH. It has been calculated that if all the heat generated by chemical reaction is absorbed by molten steel, oxidation of 1kg of Al can raise the temperature of 1t molten steel by 35℃. It is considered that the heating efficiency of aluminum and oxygen is about 60%, the calculation result is that the temperature rises 5.6℃ per minute.

钢液的铝—氧加热法（AOH）是化学热法的一种，它是利用喷枪吹氧使钢中的溶解铝氧化放出大量的化学热，从而使钢液迅速升温。该法具有许多优点：由于吹氧时喷枪浸在钢水中，所以会产生很少的烟气；由于氧气全都与钢水直接接触，所以可以准确地预测升温结果；对钢包寿命没有影响；能获得高洁净度的钢水。类似这种方法还有 CAS-OB 和 RH-OB 等。有计算表明，若化学反应产生的热量全部被钢水吸收，氧化 1kg 的 Al 可使 1t 钢水升温 35℃。考虑到铝氧加热效率约为 60%，计算结果为每分钟升温 5.6℃。

The following points should be paid attention to when refining outside the furnace with Aluminum—Oxygen heating method:

在铝—氧加热法进行炉外精炼时，需要注意以下事项：

(1) Add enough aluminum to the liquid steel, and ensure that it is completely dissolved in the steel, or float on the liquid steel. The method of adding aluminum can be realized by feeding wire, especially the aluminum wire wrapped by thin steel sheet.

(1) 向钢液中加入足够数量的铝，并保证全部溶解于钢中，或呈液态浮在钢液面上。加铝方法可通过喂线，特别是喂薄钢皮包裹的铝线。

(2) Blow enough oxygen into the liquid steel. The insertion depth and oxygen supply of the oxygen gun can be controlled quantitatively according to the needs, so that all the oxygen blew can be directly contacted with the molten steel, the utilization rate of oxygen is high, and less smoke and dust can be produced, so that the oxidation amount of aluminum and the result of temperature rise can be accurately predicted. During oxygen blowing, aluminum is oxidized first, but with the decrease of aluminum in the local area around the nozzle, silicon, manganese and other elements in the steel will also be oxidized. The oxides of silicon, manganese, iron and other elements will react with the remaining aluminum in the steel, and most of the oxides will be reduced; and some of the unreduced oxides become soot, and the other part remains in the slag. The oxygen utilization rate of this heating method is very high. Almost all of the oxygen acts directly or indirectly with aluminum, which can predict the control of aluminum content in steel more accurately.

(2) 向钢液吹入足够数量的氧气。可根据需要定量地控制氧枪插入的深度和供氧量，这样可使吹入的氧气全部直接与钢液接触，从而使得氧气利用率高，产生的烟尘少，由此可准确地预测铝的氧化量和升温的结果。吹氧期间，铝首先被氧化，但是随着喷枪口周围局部区域中铝的减少，钢中的硅、锰等其他元素也会被氧化。硅、锰、铁等元素的氧化物会与钢中剩余的铝进行反应，大多数氧化物会被还原；未被还原的氧化物一部分变成了烟尘，另一部分留在渣中。这种加热方法的氧气利用率很高，几乎全部氧气都直接或间接地与铝作用，通常可较为准确预测钢中铝含量的控制情况。

(3) In addition to controlling the amount of aluminum addition and oxygen blowing, argon is also needed to stir the molten steel for the purpose of homogenizing the temperature and composition of the molten pool, promoting the discharge of oxidation products and controlling the amount of slag from the primary smelting furnace to reduce the burning loss of aluminum.

(3) 铝氧加热除了要控制加铝量和吹氧量外，还需要使用氩气等对钢液进行搅拌，其目的是为了均匀熔池温度和成分，促进氧化产物排出的同时控制初炼炉下渣量，从而减少铝的烧损。

1.3.1.4　Other Heating Methods

1.3.1.4　其他加热方法

In addition to the above main heating methods, other methods used for heating refining molten steel include DC arc heating, electroslag heating, induction heating, plasma arc heating, electron bombardment heating, etc. These heating methods are mature in technology. When they are used in refining furnace and combined with other refining methods, there will be no insurmountable difficulties. However, these heating methods will complicate the equipment to some extent and increase the investment. They are generally used in the production of special materials and colored products.

除以上主要加热方法外，还有可以作为加热精炼钢液的其他方法，包括直流电弧加热法、电渣加热法、感应加热法、等离子弧加热法和电子轰击加热法等。这些加热方法在技术上都是成熟的，使用到精炼炉上并与其他精炼手段相配合中，也不会出现难以克服的困难。但是这些加热方法不同程度上使设备复杂化，从而增加投资。因此这些方法一般会在特种材料制作，以及有色产品中使用。

1.3.2　Selection of Heating Process

1.3.2　加热工艺的选择

The following factors should be taken into account in the correct selection of heating process for refining outside the furnace in combination with the actual situation of the plant (ladle size, characteristics of primary smelting furnace, production rhythm and steel requirements, etc.):

在正确选择炉外精炼加热工艺时，应结合工厂的实际情况（钢包大小、初炼炉特点、生产节奏和钢种要求等）考虑，重点考虑的因素包括：

(1) Heating power (energy input density in kW/T): Generally speaking, the larger the temperature, the faster the heating effect ($\dot{\varepsilon}$). However, due to the limitation of melting loss index, blowing strength, exhaust and decarburization of ladle refractories, $\dot{\varepsilon}$ can not be very high. How to further improve the heating power is worthy of further study.

(1) 加热功率（能量投入密度 $\dot{\varepsilon}$，kW/t）：一般来说，$\dot{\varepsilon}$ 越大升温越快，加热效果越好。但由于钢包耐火材料熔损指数、吹炼强度、排气量和脱碳量的限制，$\dot{\varepsilon}$ 不可能很高。如何进一步提高加热功率是值得进一步研究的课题。

(2) The larger the heating range, the more flexible the refining. Generally, the heating range of decarburization is limited by the amount of decarburization. For arc heating, due to the melting loss of furnace lining, the heating time is generally limited, which is no more than 15min, and the temperature rise range is 40~60℃.

(2) 升温幅度越大，精炼越灵活。通常情况下，脱碳加热的升温幅度受脱碳量限制。对于电弧加热，由于炉衬的熔损，一般限制的加热时间不多于15min，升温幅度为40~60℃。

(3) In order to reduce the cost, the heating range of chemical heating method should not be too large.

(3) 从降低成本出发,化学加热法的升温幅度不宜过大。

(4) The smaller the influence on the quality of molten steel, the better.

(4) 钢水质量的影响应越小越好。

Task 1.4 Vacuum
任务1.4 真空

When the reactant is a gas, the reaction can move towards the direction of generating gaseous substance by reducing the pressure of the reaction system and vacuuming. Therefore, the gas, carbon and oxygen content in the steel can be reduced by using a special vacuum device to refine the liquid steel in a vacuum environment, and the liquid steel will be further degassed, decarburized and deoxidized. In the previous study, the stirring method is used to blow argon into the molten steel. In addition to the stirring effect, each small bubble floating up from the molten steel is a 'Small Vacuum Chamber'. The partial pressures of H_2, N_2 and CO in the bubble are close to zero, and [H], [N] and carbon oxygen reaction producting CO in the steel will diffuse into the small bubble and then float up to exclude. Therefore, argon blowing has the function of 'Gas Washing' for molten steel. Common vacuum devices mainly include VD furnace, RH furnace, etc. The process of refining outside the furnace by means of vacuum is shown in Figure 1-11.

当反应生成物为气体时,通过减小反应体系的压力,采用抽真空的手段,可以使反应向着生成气态物质的方向移动。因此,在真空下,采用专门的真空装置,将钢液置于真空环境中精炼,可以降低钢中气体、碳及氧含量,钢液将进一步脱气、脱碳和脱氧。在之前的学习搅拌手段中介绍向钢液中吹入氩气,除了有搅拌作用之外,吹入的氩气从钢液中上浮的每个小气泡都相当一个"小真空室",气泡内的H_2、N_2和CO等分压接近于零,钢中的[H]、[N]和碳氧反应产物CO将向小气泡中扩散并随之上浮排除。因此,吹氩对钢液具有"气洗"作用。常用的真空装置主要有VD炉、RH炉等。以真空为手段进行炉外精炼的工艺流程如图1-11所示。

When the metallurgical reaction product is gas, the reaction equilibrium can move towards the direction of gas-phase material by reducing the pressure of the reaction system (i.e. vacuuming). Therefore, under vacuum, the liquid steel will be further degassed, decarburized and deoxidized. For example, it is difficult to reduce [C] to a very low value at 1873K for the return oxygen blowing method of EAF smelting stainless steel. In the AOD method, the co partial pressure produced in the carbon oxygen reaction can be reduced by blowing the gas continuously changing the Ar/O_2 ratio into the molten steel, so that [C] content of the molten steel can reach the ultra-low carbon level. The non-ferrous metals in molten steel (such as Pb, Cu, as, Sn and Bi), can also be removed by volatilization during the vacuum melting or treatment of molten steel. The amount of volatilization depends on the vapor pressure of the element and its activity in the molten iron.

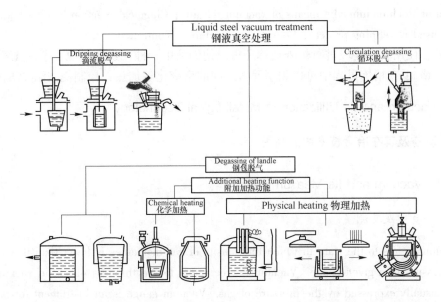

Figure 1-11 Secondary refining process with vacuum as the main means
图 1-11 以真空为主要手段的炉外精炼工艺流程

当冶金反应生成物为气体时，通过减小反应体系的压力（即抽真空），可以使反应的平衡向着生成气态物质的方向移动。因此，在真空下，钢液将进一步脱气、脱碳和脱氧。例如，电弧炉冶炼不锈钢的返回吹氧法，在1873K下很难使 [C] 的数值降至很低。而在 AOD 法中，向钢液中吹入不断变换 Ar/O_2 比例的气体，可以降低碳氧反应中产生的 CO 分压，从而使钢液的 [C] 含量达到超低碳水平。钢液中的有色金属（如 Pb、Cu、As、Sn、Bi 等）在钢液进行真空熔炼或处理时，也可通过挥发除去一部分金属。其挥发量取决于该元素的蒸汽压和在铁液中的活度。

Therefore, the purposes of using vacuum as a means in the secondary refining are as follows:
因此，在炉外精炼中采用真空作为手段的目的包括：

(1) Remove hydrogen and oxygen, and reduce nitrogen content to a lower range.

(1) 脱除氢和氧，并将氮气含量降至较低范围。

(2) Remove non-metallic inclusions to improve the cleanliness of molten steel.

(2) 去除非金属夹杂物，从而改善钢水的清洁度。

(3) Production of ultra-low carbon steel (The carbon content of ultra-low carbon steel does not have a recognized strict standard. In recent years, it is considered that the steel with $w[C]<0.015\%$, or even $w[C]<0.005\%$ is ultra-low carbon steel); The one element oxidized is prior to other elements (such as carbon prior to chromium); The chemical is heated; The casting temperature is controlled, etc.

(3) 生产超低碳钢（超低碳钢的碳含量没有一个公认的严格标准，近年来认为$w[C]<0.015\%$，甚至$w[C]<0.005\%$的钢种均视为超低碳钢）；使一种元素比其他元素优先氧化（如碳优先于铬）；化学加热；控制浇铸温度等。

In addition, the one element in molten steel can be oxidized prior to other elements (such as carbon prior to chromium) by means of vacuum refining; Chemical reaction heating between elements is used in refining process; Pouring temperature is controlled, etc.

另外，使用真空手段进行炉外精炼还可以使钢液中一种元素比其他元素优先氧化（如碳优先于铬）；在精炼过程中可以利用元素之间的反应化学加热；控制浇注温度等。

1.4.1 Vacuum and Its Utilization in Metallurgical Technology

1.4.1 真空及其在冶金技术中的利用

1.4.1.1 Vacuum and Its Measurement

1.4.1.1 真空及其测量

Vacuum refers to the state in which the density of gas molecules is lower than that of atmospheric pressure in a given space. Vacuum degree refers to the thinness of gas in vacuum state, which is usually expressed by the pressure of gas. Vacuum gauge is an instrument for measuring vacuum degree and a measuring tool for vacuum degree. The commonly used vacuum gauge and its measuring range are shown in Table 1-12. The following methods are generally adopted in the international vacuum area division as follows:

真空是指在给定的空间内，气体分子的密度低于该地区大气压的气体分子密度的状态。真空度是指处于真空状态下的气体的稀薄程度，常用气体的压强来表示。真空计是测量真空度的仪器及其真空度的测量工具，常用真空计及其测量范围见表1-12。真空区域的划分国际上通常采用如下办法：

(1) Coarse vacuum（粗真空）$<(760\sim1)\times133.3$ Pa.
(2) Medium vacuum（中真空）$<(1\sim10^{-3})\times133.3$ Pa.
(3) High vacuum（高真空）$<(10^{-3}\sim10^{-7})\times133.3$ Pa.
(4) Ultra high vacuum（超高真空）$<10^{-7}\times133.3$ Pa.

Table 1-12 Measurement range of various vacuum meters
表1-12 各种真空计的测量范围

Name of vacuum gauge 真空计名称	Measuring range/Pa 测量范围/Pa
Mercury manometer 水银压力计	$(1\sim760)\times133.3$
Oil pressure gauge 油压力计	$(0.02\sim20)\times133.3$
Maxwell meter 麦氏计	$(0\sim10^{-5})\times133.3$
Clarinet vacuum gauge 单簧管真空计	$(10\sim760)\times133.3$
Diaphragm vacuum gauge 隔膜真空计	$(10^{-4}\sim10)\times133.3$

Continued Table 1-12

Name of vacuum gauge 真空计名称	Measuring range/Pa 测量范围/Pa
Resistance vacuum gauge 电阻真空计	$(10^{-4} \sim 100) \times 133.3$
Thermocouple vacuum gauge 热偶真空计	$(10^{-3} \sim 1) \times 133.3$

At present, the working vacuum degree of the refining equipment outside the furnace can be dozens of PA, and its ultimate vacuum degree should have the ability to reach about 20Pa. As far as the vacuum treatment of liquid steel is concerned, it belongs to the low vacuum area and is generally measured by U-tube and compression vacuum gauge.

现在炉外精炼设备的工作真空度可以在几十帕,而其极限真空度应该具有达到20Pa左右的能力。钢液真空属于低真空区域,一般使用U形管和压缩式真空计来测量。

1.4.1.2 Overview of Vacuum Metallurgy

1.4.1.2 真空冶金概述

Vacuum metallurgy is different from the metallurgical process in the atmosphere. It is easy to oxidize the active metal, and the alloy composition is difficult to control accurately. When the metal bath and the air act, the harmful gases (H, N, O) in the alloy are difficult to remove. The smelting in the atmosphere suppresses the volatilization process, and can not remove the harmful elements Pb, Sb, Bi, Sn, Cu, etc.

真空冶金不同于大气下的冶金过程,对诸如活泼金属易于氧化,合金成分难精确控制。金属熔池与空气作用,合金中有害气体(H、N、O)难去除。大气下熔炼抑制了挥发过程,不能去除有害元素Pb、Sb、Bi、Sn、Cu等,都有较好的精炼作用。

In a broad sense, vacuum metallurgy includes vacuum degassing, vacuum melting, vacuum casting, vacuum distillation, vacuum decomposition, vacuum sintering, vacuum heat treatment, vacuum welding and vacuum coating.

广义上真空冶金包括真空脱气、真空熔炼、真空熔铸、真空蒸馏、真空分解、真空烧结、真空热处理、真空焊接和真空镀膜等。

Vacuum pump is often used to create vacuum state in metallurgical production. Vacuum pump can be basically divided into two categories: gas delivery pump and gas collection pump. The main functions of vacuum pump are to form a limit vacuum state, to reasonably regulate the pumping speed and air volume, and to make full use of the maximum back pressure to create a vacuum state.

冶金生产中常使用真空泵去创造真空状态,真空泵基本上可以分为两大类,即气体输送泵和气体收集式泵。真空泵的主要性能包括:能够形成极限真空状态;合理调节抽气速度及抽气量;充分利用最大反压强创造真空状态。

The working process of the vacuum jet pump is divided into three stages: The first stage is the expansion of the working steam in the nozzle; The second stage is the mixing of the working steam with the extracted gas in the mixing chamber; The third stage is the compression of the mixed gas in the diffuser.

真空喷射泵的工作过程为三个阶段：第一个阶段为工作蒸汽在喷嘴中膨胀；第二阶段为工作蒸汽在混合室中与被抽气体混合；第三阶段为混合气体在扩压器中被压缩。

The metallurgical functions that can be realized by using vacuum include: the dissolution and precipitation of gas in molten steel; the deoxidation and decarbonization reaction of molten steel is controlled; The effect of carbon in molten steel is dissolved in molten steel on furnace lining; The volatilization of alloy elements is controlled; The volatilization of metal and non-metal inclusions is removed.

利用真空可以实现的冶金功能有气体在钢液中的溶解和析出；控制钢液的脱氧及脱碳反应；使钢液或溶解在钢液中的碳与炉衬的作用；控制合金元素的挥发；去除金属夹杂及非金属夹杂的挥发。

1.4.2 Vacuum Degassing of Liquid Steel

1.4.2 钢液的真空脱气

Gas in steel refers to hydrogen and nitrogen dissolved in steel. The gas sources include metal material, slag making agent, refractory, liquid steel suction, oxygen for steelmaking, etc. The solubility of gas in steel increases with the increase of temperature. When making steel, try to avoid high temperature tapping. In addition, the pressure, composition, phase transformation and alloy elements in the molten steel smelting process will also affect the gas content in the molten steel. For example, the higher the partial pressure of H_2 and N_2 in gas phase, the higher the solubility of gas in steel, so the dehydrogenation effect under vacuum is better, and the denitrification effect is not ideal. If the activity coefficient of gas in steel is small, the solubility is high. The gas content in the molten steel can be reduced by reducing the partial pressure. The solubility of gas in solid pure iron is lower than that in liquid iron. α-Fe changes to γ-Fe at 910℃, γ-Fe changes to α-Fe at 1400℃, and the solubility also changes abruptly. The effects of alloy elements on gas in molten steel are shown in Tables 1-13 and 1-14.

钢中气体是指溶解在钢中的氢和氮，其气体来源包括金属料、造渣剂、耐火材料、钢液吸气、炼钢用氧气等。随着钢液冶炼过程温度升高，气体在钢中的溶解度提高。炼钢时，要尽量避免高温出钢。另外钢液冶炼过程的压力、成分、相变、合金元素也会影响钢液中气体的含量。例如，气相中 H_2、N_2 的分压越大，气体在钢中的溶解度越高，真空下的脱氢效果越好，但脱氮效果越不理想。如果气体在钢中的活度系数小，则溶解度高。钢液中的气体可通过降低分压来降低含量。固态纯铁中气体的溶解度低于液态纯铁中气体的溶解度。910℃时发生 α-Fe 向 γ-Fe 转变，1400℃时发生 γ-Fe 向 α-Fe 转变，溶解度也发生突变。合金元素对钢液中气体的影响见表1-13和表1-14所示。

Table 1-13 Effects of alloy elements on hydrogen activity coefficient and solubility
表 1-13 合金元素对氢活度系数和溶解度的影响

Alloying element 合金元素	Influence on f_H and $w[H]$ 对 f_H 和 $w[H]$ 的影响
Ti, V, Cr, Nb	$w[H]$ Decrease f_H and increase $w[H]$ 降低 f_H，增加 $w[H]$
C, Si, B, Al	Decrease f_H and increase $w[H]$ 增加 f_H，降低 $w[H]$
Mn, Co, Ni, Mo	No effect 无影响

Table 1-14 Effects of alloy elements on nitrogen activity coefficient and solubility
表 1-14 合金元素对氮活度系数和溶解度的影响

Alloying element 合金元素	Influence on f_N and $w[N]$ 对 f_N 和 $w[N]$ 的影响
V, Nb, Cr, Ti	Decrease f_N and increase $w[N]$ 降低 f_N，增加 $w[N]$
Mn, Mo, W	Little effect on f_N and $w[N]$ 对 f_N 和 $w[N]$ 影响较小
C, Si	Significantly increase f_N and decrease $w[N]$ 增加 f_N，减少 $w[N]$

1.4.2.1 Hazards of Gas in Molten Steel

1.4.2.1 钢液中气体的危害

The presence of gas can significantly reduce the properties of steel, and it is easy to cause many defects of steel. When the molten steel solidifies, the solubility decreases sharply. For example, when the temperature of molten steel drops from 1650℃ to 410℃, the solubility of hydrogen drops to 1/83 of the original value. Because hydrogen is solid dissolved in the steel in the form of atoms or ions, forming a gap solid solution gap. After solidification, the atomic spacing in solid iron is much closer than that in liquid steel, resulting in the decrease of solubility. The gas segregation occurs in the solidification process, and the gas is concentrated in the center of the ingot (billet), and the formation of the central pore or micro pore is promoted by the increase of the depth. Hydrogen can cause white spots (cracks) in steel. When the content of nitrogen increases, the strength of the steel increases, but the plasticity and weldability decrease. At the same time, the cold brittleness and age hardening of the steel occur. Therefore, the steel with good toughness should reduce the nitrogen content.

气体的存在显著降低钢的性能，而且容易造成钢的许多缺陷。当钢液凝固时，溶解度急剧减小，例如，钢液温度从1650℃降至410℃时，氢的溶解度降至原来的1/83。这是因为氢在钢液内以原子或离子形式固溶于钢中形成间隙固溶体，凝固后的固体铁中原子的间距比液态时紧密得多，从而造成溶解度下降，致使气体在凝固过程中产生偏析现象，聚集到钢锭（坯）的中心部位，深度逐渐增大，促成中心孔隙或显微孔隙的形成。氢能使钢产生白点（发裂）。氮含量增加时，钢的强度增强，但塑性和焊接性能降低。同时引起钢的冷脆性，产生时效硬化。因此，要求韧性良好的钢降低氮的含量。

1.4.2.2 Vacuum Degassing Principle

1.4.2.2 真空脱气原理

The gas content dissolved in steel is mainly determined by the partial pressure of water vapor and nitrogen in the furnace gas. The partial pressure p_{N_2} of nitrogen in the furnace gas is kept at 0.79×10^5 Pa, while the pH in the furnace gas is very low, about 5.35×10^{-2} Pa. Therefore, the hydrogen in steel is mainly determined by the partial pressure of water vapor in the furnace gas.

溶解于钢中的气体含量主要决定于炉气中水汽和氮气的分压力。氮气在炉气中的分压力 p_{N_2} 大体上保持在 0.79×10^5 Pa，而炉气中的pH很低，约 5.35×10^{-2} Pa，因此钢中的氢主要由炉气中水蒸气的分压决定。

Hydrogen and nitrogen have certain solubility in various states of steel, and the solubility increases with the increase of temperature due to the endothermic process. When gaseous hydrogen and nitrogen are dissolved in pure iron or liquid steel, the gas molecules are first adsorbed on the Gas—Liquid interface and decomposed into two atoms, which are then absorbed by the liquid steel. During vacuum degassing, the gas dissolved in the molten steel is discharged due to the reduction of the partial pressure of the gas phase.

氢和氮在各种状态的钢中都具有一定的溶解度。由于溶解过程吸热，故溶解度随温度的升高而增加。气态的氢和氮在纯铁或钢液中溶解时，气体分子先被吸附在气—液界面上，并分解成两个原子，然后这些原子被钢液吸收。真空脱气时，因降低了气相分压力而使溶解在钢液中的气体排出。

The interaction between metal and diatomic gas is similar to chemical reaction, and the number of gas moles changes during the interaction. Therefore, the change of gas partial pressure at a certain temperature will cause the equilibrium to move and change the gas content in the liquid metal. Vacuum degassing is based on this principle. Therefore, a certain degree of vacuum must be created to achieve a certain degassing effect.

金属与双原子气体的相互作用与化学反应相似，作用过程中气体的摩尔数有变化。因此，当温度一定时气体分压力的变化将引起平衡移动，使金属液内的气体含量发生变化。真空脱气就是根据这个道理进行的，所以要达到一定的脱气效果，就得造成一定的真空度。

It can be seen from the above that in order to reduce the hydrogen and nitrogen content in the steel to a small amount, it is not necessary to maintain a high vacuum, which can be achieved at a pressure of several hundred. However, the actual degassing effect is also limited by dynamic factors.

由上可见，要把钢中的氢和氮降低到较少的含量，并不需要保持很高的真空度，几百帕的压力下就可以实现。但是，实际的脱气效果还要受动力学因素的限制。

During vacuum degassing, the gas dissolved in the molten steel is discharged due to the reduction of the partial pressure of the gas phase. From the thermodynamic point of view, when the partial pressure of hydrogen and nitrogen in the gas phase is 100~200Pa, the gas content can be reduced to a lower level.

真空脱气时，因降低了气相分压力，才使溶解在钢液中的气体排出。从热力学角度分析，当气相中氢和氮的分压力为100~200Pa时就能将气体含量降到较低水平。

The dissolution of nitrogen and hydrogen in molten steel follows the law of sewart (square root). The pressure of the system is reduced, and the partial pressure of the gas is reduced, so as to reduce the amount of dissolved gas in the molten steel.

氮、氢在钢液中的溶解遵从西华特（平方根）定律，降低体系的压力，使气体的分压降低，从而减小钢液中溶解的气体量。

The dynamic degassing reaction steps of vacuum degassing are as follows：

真空脱气的动力学脱气反应步骤如下所示：

(1) Through convection or diffusion (or the combination of the two), the gas atoms dissolved in the molten steel migrate to the liquid steel of Liquid-Gas interface.

(1) 通过对流或扩散（或两者的综合），溶解在钢液中的气体原子迁移到钢液的液—气相界面。

(2) The gas atom changes from dissolved state to adsorbed state.

(2) 气体原子由溶解状态转变为表面吸附状态。

(3) The gas atoms adsorbed on the surface interact with each other to form gas molecules.

(3) 表面吸附的气体原子彼此相互作用，生成气体分子。

(4) Gas molecules are desorbed from the surface of molten steel.

(4) 气体分子从钢液表面脱附。

(5) Gas molecules diffuse into the gas phase and are pumped out by a vacuum pump.

(5) 气体分子扩散进入气相，并被真空泵抽出。

At the steel-making temperature, the steps of (2), (3), (4) and (5) are carried out very fast. In the vacuum with the gas phase pressure less than atmospheric pressure, the diffusion speed of gas molecules is also very fast. Therefore, the speed of vacuum degassing depends on the migration of gas dissolved in steel to the Gas—Liquid interface in the step of (1). In various vacuum degassing methods, there are different forms of agitation measures, so the gas transfer in the molten steel is also very fast. The limiting link to control degassing speed is the mass transfer speed of gas atoms through the diffusion boundary layer of molten steel. Because of the limited surface area, the mass transfer rate of gas through the boundary layer is very small.

在炼钢温度下，步骤（2）、（3）、（4）和（5）的进行速度相当快，在气相压力小于大气压的真空中，气体分子的扩散速度相当迅速，所以真空脱气的速度取决于步骤（1），溶解在钢中的气体向气—液相界面迁移。在各种真空脱气方法中都配有不同形式的搅拌措施，所以气体在钢液内传递迅速，控制脱气速度的限制性环节是气体原子通过钢液扩散边界层的传质速度。由于表面积有限，所以气体通过边界层的传质速度较小。

Therefore, there are three steps in the degassing process of liquid steel in vacuum：

因此，真空中钢液脱气过程包括以下三个环节：

(1) The dissolved gas atom in molten steel diffuses to the liquid steel of Liquid—Gas interface, and the gas atom passes through the Gas—Liquid interface layer.

(1) 钢液中溶解气体原子向钢液的液—气相界面扩散，气体原子通过气—液界面层。

(2) These gas atoms are adsorbed on the phase interface, combined into gas molecules, and then desorbed from the interface.

(2) 这些气体原子在相界面上吸附，结合成气体分子，再从界面脱附。

(3) The desorbed gas molecules diffuse into the gas phase under vacuum and are pumped out by vacuum pump.

(3) 脱附的气体分子在真空作用下向气相中扩散，并被真空泵抽出。

Among them, the step of (1) is a restrictive link.

其中，步骤 (1) 为限制性环节。

The limiting link of degassing process is that the gas dissolved in steel passes through the boundary layer on the liquid steel side of the steel gas interface, so the gas diffusion rate in the boundary layer on the liquid steel side can be regarded as the total rate of degassing process.

真空脱气速率脱气过程的限制性环节是溶解于钢中的气体穿过钢—气相界面钢液侧的边界层，所以钢液侧边界层中气体的扩散速率可以当作脱气过程的总速率。

Because the limiting link of degassing process is the migration of gas dissolved in steel to the Gas—Liquid interface, the mass transfer rate of gas atoms from the diffusion boundary layer to the Gas—Liquid interface can be regarded as the total speed of degassing process. The degassing rate is fast when the molten pool is boiling and stirring with argon blowing.

因为脱气过程的限制性环节是溶解在钢中的气体向气—液相界面迁移，所以气体原子通过钢液扩散边界层向气—液相界面的传质速度可以当作脱气过程的总速度。在熔池沸腾和吹氩搅拌时，脱气的速率快。

At first, the partial pressure of hydrogen and nitrogen in the argon bubble and carbon monoxide bubble is zero. The dissolved hydrogen and nitrogen gradually diffuse from the molten steel to the bubble, and automatically balance the chemical potential of hydrogen and nitrogen in the two phases. Finally, carbon monoxide and argon bubbles float out of the liquid surface and enter the gas phase. The more bubbles are produced in unit time, the longer bubbles stay in molten steel, the greater gas mass transfer coefficient, the greater difference between actual concentration and equilibrium concentration, and the greater degassing speed.

吹入氩气泡和碳—氧反应产生的一氧化碳气泡对脱氢和脱氮有促进作用，起初在氩气泡和一氧化碳气泡内氢和氮的分压力为零。溶解的氢和氮逐步从钢液内向气泡内扩散，自动地在两相中使氢和氮的化学位实现平衡。最后借助一氧化碳和氩气泡浮出液面而进入气相。在处理过程中，单位时间内产生的气泡数越多，气泡在钢液内停留时间越长，气体传质系数越大，实际浓度和平衡浓度之差越大，则脱气速度越大。

1.4.2.3 Classification of Vacuum Degassing

1.4.2.3 真空脱气分类

According to the characteristics of steelmaking process and the equipment used in the refining workshop outside the furnace, there are three kinds of vacuum degassing widely used at present:

根据炼钢工艺特点及生产现场炉外精炼车间的使用设备，目前广泛使用的真空脱气有以下三种：

(1) Steel flow degassing: The falling steel flow is exposed to vacuum and then collected into ingot mould, ladle or furnace.

(1) 钢流脱气：下落中的钢流被暴露给真空，然后被收集到钢锭模、钢包或炉内。

(2) Ladle degassing: The molten steel in the ladle is exposed to vacuum and stirred with gas or electromagnetic.

(2) 钢包脱气：钢包内钢水被暴露给真空，并用气体或电磁搅拌。

(3) Circulation degassing: The molten steel in the ladle is compressed into the evacuated vacuum chamber by atmospheric pressure, exposed to vacuum, and then flows out of the degassing chamber into the ladle.

(3) 循环脱气：在钢包内的钢水由大气压力压入抽空的真空室内，暴露给真空，然后流出脱气室进入钢包。

1.4.2.4 Measures to Reduce Gas in Steel

1.4.2.4 降低钢中气体的措施

According to the principle of vacuum degassing, the following measures are taken to reduce the gas content in the molten steel during the refining process outside the furnace:

根据真空脱气的原理，炉外精炼生产过程中降低钢液中气体含量的措施包括：

(1) Dry raw materials and refractories are used.

(1) 使用干燥的原材料和耐火材料。

(2) To reduce the partial pressure of gas in the gas phase contacting with liquid steel, it can be done in two ways. One is to reduce the total pressure of gas phase, that is, to use vacuum degassing to keep the liquid steel in a low pressure environment. Various measures can also be taken to reduce the static pressure caused by molten steel and slag. On the other hand, the partial pressure is reduced by dilution. For example, the partial pressure is very low in the bubbles formed by the reaction of argon blowing and carbon oxygen to produce carbon monoxide gas.

(2) 可从两方面采取措施降低与钢液接触的气相中气体的分压。一是降低气相的总压，即采用真空脱气，将钢液处于低压的环境中。也可采用各种减小钢液和炉渣所造成的静压力的措施。另一方面是用稀释的办法来减小分压，如吹氩、碳氧反应产生一氧化碳气体所形成的气泡中，分压极低。

(3) The specific surface area of liquid steel during degassing (A/V) is increased. On the one

hand, it can reduce the partial pressure of hydrogen and nitrogen in the gas phase. On the other hand, it can also increase the diffusion mass transfer coefficient and unit degassing area.

（3）在脱气过程中增加钢液的比表面积（A/V）。增大吹氩搅拌的氩气流量，一方面能降低氢和氮在气相中的分压，另一方面也增大了脱气过程的扩散传质系数和单位脱气面积。

（4）The mass transfer coefficient is improved.

（4）提高传质系数。

（5）The degassing time as appropriate is properd.

（5）适当地延长脱气时间。

（6）The nitrogen is removed by the generated nitride.

（6）利用生成的氮化物去除氮。

（7）When other operations are not affected, the tapping temperature shall be reduced as much as possible to reduce the solubility of gas in the steel.

（7）不影响其他操作时，尽量降低出钢温度，减小气体在钢中的溶解度。

（8）In smelting, the pool boiling produced by decarburization should be fully utilized to reduce the gas content in molten steel.

（8）冶炼中，应充分利用脱碳反应产生的溶池沸腾来降低钢水中的气体含量。

It has been proved that hydrogen removal is better, and no matter what vacuum treatment is adopted, good dehydrogenation effect can be maintained. However, the denitrification effect of various degassing methods is not ideal. The reasons are as follows：

实践证明，不管采用哪种真空处理都能保持良好的脱氢效果。但是各种脱气方法的脱氮效果都不甚理想。其原因归纳如下：

（1）The diffusion coefficient of nitrogen is smaller than that of hydrogen. The nitrogen in steel mainly exists in the form of nitride. The nitrogen in free state is less, and the nitrogen in free state can also diffuse. Nitrogen is two orders of magnitude larger than that of hydrogen atom, so the diffusion speed of nitrogen is much smaller than that of hydrogen. The solubility of nitrogen in steel is 15 times higher than that of hydrogen.

（1）氮的扩散系数比氢小，钢中的氮主要呈氮化物状态存在，自由状态的氮较少，并且自由状态的氮也能扩散。氮比氢原子大两个数量级，因此氮的扩散速度比氢小得多。氮在钢中的溶解度比氢高15倍。

（2）Nitrogen reacts with some elements to form nitrides. In order to remove nitrogen from nitrides, the gas pressure (vacuum degree) must be less than the decomposition pressure of nitrides, which are relatively low at 1600℃.

（2）氮与某些元素作用生成氮化物。若要脱除氮化物中的氮，则气相压力（真空度）必须小于氮化物的分解压，而这些氮化物在1600℃下的分解压都比较低。

（3）The existence of active elements oxygen and sulfur on the surface of molten steel occupied the surface of molten steel, and the chemical reaction of denitrification was blocked. Therefore, denitrification is based on deoxidation and desulfurization.

（3）钢液表面活性元素氧、硫占据钢液表面，脱氮化学反应受阻。因此，脱氮以脱氧和脱硫为前提。

(4) The atmosphere contains 78% nitrogen (by volume), and the liquid steel will absorb nitrogen as long as it contacts with the atmosphere.

(4) 大气中含氮气78%（体积分数），因此钢液只要和大气接触就会吸氮。

Some scholars have studied the use of synthetic slag for denitrification. The synthetic slag has a very high solubility for nitrogen under reduction conditions, with a solubility of 1.33% ~ 1.88%, but the effect is still under study.

有学者研究使用合成渣脱氮，合成渣在还原条件下对氮有非常高的溶解能力，其溶解度为1.33%~1.88%，但效果还在研究中。

1.4.3　Vacuum Deoxidation of Molten Steel

1.4.3　钢液的真空脱氧

At atmospheric pressure, the deoxidization ability of carbon is very weak. So strong deoxidizers (such as silicon and aluminum) must be used for final deoxidization, but the deoxidization ability of carbon under vacuum is significantly enhanced. As the deoxidation product of carbon is CO gas, the equilibrium moves towards the direction of CO formation under the condition of reducing pressure, which increases the deoxidization ability of carbon. The deoxidization ability of carbon in vacuum is equal to that of silicon and aluminum.

在常压下碳的脱氧能力很弱，必须使用强脱氧剂（如硅、铝）进行终脱氧，但在真空下碳的脱氧能力显著增强。由于碳的脱氧产物是CO气体，在降低压力的情况下，平衡向生成CO的方向移动，使碳的脱氧能力增大。在真空下碳的脱氧能力达到了和硅、铝相当的水平。

1.4.3.1　Carbon Oxygen Reaction Mechanism

1.4.3.1　碳氧反应机理

Vacuum decarburization and deoxidation are carried out at the same time: [C]+[O]=CO under vacuum condition, the partial pressure of CO is reduced, and the decarburization reaction which has reached equilibrium under atmospheric pressure is carried out again, so as to achieve the purpose of decarburization and deoxidation.

真空脱碳和脱氧同时进行：[C]+[O]=CO在真空条件下，降低了CO分压，使在大气压力下已经达到平衡的脱碳反应再度进行，从而达到脱碳脱氧的目的。

The deep decarburization of molten steel can be achieved by using vacuum, and the oxygen content in molten steel can be reduced to a very low level. Under vacuum condition, the deoxidizer is carbon. When the concentration of deoxidizing element [B] = 0.1%, p_{CO} = 10kPa, the deoxidizing ability of carbon is higher than that of silicon; while when p_{CO} = 0.1kPa, the deoxidizing ability of carbon is even higher than that of aluminum. The limiting link of the rate of vacuum deoxidation process is the diffusion of [C] and [O] in molten steel. When [C] is high, the diffusion of [O] is the limiting link. The kinetics of decarburization shows that the vacuum decarburization can be used when the carbon content in the molten steel decreases below the critical value ([C]<0.1%~0.45%).

利用真空可使钢液深度脱碳,生产超低碳钢时使钢液中氧含量降低到很低的水平。真空条件下,脱氧剂为碳,当脱氧元素浓度[B]=0.1%, p_{CO} =10kPa 时,碳的脱氧能力高于硅的脱氧能力;当 p_{CO} =0.1kPa 时,碳的脱氧能力甚至高于铝的脱氧能力。真空脱氧过程速率的限制性环节是钢液中[C]和[O]的扩散,[C]较高时,[O]的扩散为限制性环节。脱碳反应动力学表明,当钢液中碳含量降低到临界量以下([C]<0.1%~0.45%后),即可采用真空脱碳。

The thermodynamic equilibrium of carbon and oxygen in vacuum is only valid at the Gas-Liquid interface. At the Gas-Liquid interface, the deoxidation product CO can be removed from the liquid surface to the gas. At this time, the equilibrium of the reaction is affected by the partial pressure of CO in the gas phase. However, in the molten steel, due to the static pressure, the pressure in CO bubble must be much higher than the partial pressure of CO in the gas phase on the metal surface. Because it is necessary to overcome the effects of total gas pressure, capillary pressure and hydrostatic pressure.

真空下碳氧热力学平衡关系只在气—液相交界面上才有效。在气—液相界面上脱氧产物 CO 能从液面上进入气体中,此时反应的平衡受气相中 CO 分压力的影响。但在钢液内部,由于静压力等作用,CO 气泡内的压力必然大大超过金属液面上气相中 CO 的分压。因为生成气泡要克服气相总压力、毛细管压力和液体静压力作用。

A lot of small bubbles are formed after blowing inert gas into the molten steel. The content of CO in these small bubbles is very small. The carbon and oxygen in the molten steel can combine to form CO on the surface of the bubbles and enter the bubbles. Until the partial pressure of CO in the bubble reaches the value of equilibrium with the phases of the content of [C] and [O] by mass in the molten steel. This is the theoretical basis for degassing and deoxidizing by argon blowing. Under vacuum condition, in various methods of stirring and circulating molten steel by blowing argon in refined molten steel, argon bubble not only plays the role of stirring and circulating, but also has the function of degassing and deoxidizing, as shown in Figure 1-12.

向钢液吹入惰性气体后形成很多小气泡,这些小气泡内的 CO 含量很少,钢液中的碳和氧能在气泡表面结合成 CO 而进入气泡内。直到气泡中的 CO 分压达到与钢液中的[C]含量、[O]含量相平衡的数值为止。这就是吹氩脱气和脱氧的理论根据。在真空条件下,在精炼的钢液中吹氩搅拌和吹氩循环钢液的各种方法中,氩气泡除起搅拌和循环作用外,同时还有这种脱气和脱氧作用,如图 1-12 所示。

The refractory surface of furnace bottom and furnace wall is rough and uneven, and the contact between molten steel and refractory is non wettable (the wetting angle is between 100° and 120°). There are always many tiny cracks and pits on the rough surface. When the cracks are very small, due to the action of surface tension, the metal can not enter, and these cracks and pits become the germination point of CO bubbles.

炉底和炉壁的耐火材料表面是粗糙不平的,钢液与耐火材料的接触是非润湿性的(润

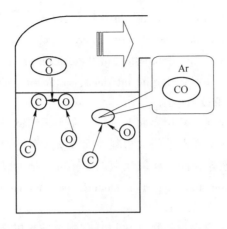

Figure 1-12 Mechanism of carbon oxygen reaction in vacuum
图 1-12 真空下碳氧反应机理

湿角在 100°~120°之间),在粗糙的表面上存在不少微小的缝隙和凹坑。当缝隙很小时,由于表面张力的作用,金属不能进入,这些缝隙和凹坑成为 CO 气泡的萌芽点。

From the formation of bubble core, until the hemisphere, the partial pressure in equilibrium with [C] and [O] must be greater than the pressure in the bubble attached to the furnace bottom or furnace wall to make the bubble continue to develop until the separation floats away.

从气泡核形成逐渐鼓起,直到半球前,[C] 和 [O] 平衡的分压力必须大于附着在炉底或炉壁上的气泡内的压力,才能使气泡继续发展直到分离浮去。

Under vacuum, the carbon oxygen reaction is only effective at the Gas-Liquid interface. The farther away the carbon oxygen reaction is from the Gas-Liquid interface, the greater the necessary r value is. In the process of deoxidization, as the value of the content of O and C is getting lower and higher, more and more small cracks and pits can no longer play the role of germination. As a result, the depth of bubbles is getting smaller with the decrease of the content of [C] and [O] by mass.

真空下碳氧反应只在气—液相界面上有效,碳氧反应离气—液相界面位置越远,r 值越大。在脱氧过程中,由于碳含量和氧含量越来越低,r 值越来越大,所以越来越多的小缝隙和凹坑不能再起萌芽作用。因此随着碳含量和氧含量的降低,产生气泡的深度越来越小。

During the vacuum treatment of molten steel, start the vacuum pump to reduce the system pressure and make the reaction balance move. When the molten steel is boiling, a large number of bubbles are generated (the highest peak), and then the bubbles are stopped on the lower wall. And the boiling is gradually weakened, which is the principle of the generation and stop of molten steel boiling in the process of carbon deoxidization under vacuum.

在真空处理钢液时,启动真空泵降低系统压力使反应平衡移动。钢液形成沸腾,大量气泡产生(最高峰),然后由于下部器壁停止生成气泡,沸腾又逐渐减弱,这就是在真空下碳脱氧过程中钢液沸腾的产生和停止原理。

The carbon oxygen reaction can only be carried out on the Gas-Liquid interface. Under the actual steelmaking conditions, the Gas-Liquid interface can be provided by the unsmooth refractory surface contacting with the liquid steel or the gas blown into the liquid steel. It can be considered that there is always a Gas-Liquid interface, so the steps of carbon oxygen reaction can be divided into the following four steps:

碳氧反应只能在现成的气—液相界面上进行。在实际炼钢条件下,这种现成的气—液相界面可以由与钢液接触的不光滑的耐火材料表面或吹入钢液内的气体来提供,因此可以认为总是存在现成的气—液相界面。碳氧反应的步骤可以分为以下四步:

(1) The dissolved carbon and oxygen in molten steel migrate to the Gas-Liquid interface through the diffusion boundary layer.

(1) 钢液内溶解的碳、氧通过扩散边界层迁移到气液相界面。

(2) CO gas is produced by chemical reaction at the Gas-Liquid interface.

(2) 在气液相界面上进行化学反应生成 CO 气体。

(3) The reaction product (CO) separated from the phase interface and entered the gas phase.

(3) 反应产物(CO)脱离相界面进入气相。

(4) CO bubbles grow up and float up and they are discharged through liquid steel.

(4) CO 气泡长大和上浮,并通过钢液排出。

Steps of (2), (3) and (4) are all carried out very fast, so the mass transfer rate of carbon and oxygen dissolved in the molten steel to the Gas—Liquid interface is the limiting link to control the reaction rate of carbon and oxygen. The diffusion coefficient of carbon in molten steel is larger than that of oxygen, and the content of carbon in molten steel is one or two orders of magnitude higher than that of oxygen. Therefore, the mass transfer of oxygen in high carbon region to the boundary layer is the limiting link of the reaction rate of carbon deoxidation under vacuum. The low carbon domain is the opposite.

步骤(2)、(3)、(4)进行的都很快,所以溶解在钢液内的碳氧向气—液相界面的传质速度是控制碳氧反应速度的限制性环节。碳在钢液内的扩散系数比氧大,钢液内碳的含量又比氧含量高一至两个数量级。因此,高碳域氧向边界层的传质是真空下碳脱氧反应速度的限制性环节,低碳域则相反。

1.4.3.2　Effective Measures for Vacuum Deoxidation

1.4.3.2　真空脱氧的有效措施

The carbon monoxide bubbles formed by the reaction of carbon and oxygen in the smelting process are non spontaneous nucleation. After blowing inert gas into the molten steel, many small bubbles are formed. The carbon monoxide content in these small bubbles is very small. The carbon and oxygen in the molten steel can combine to form carbon monoxide on the surface of the bubbles and enter the bubbles. In addition, the surface of refractories at the bottom and wall of the furnace is rough, with small gaps and pits, which also create conditions for the formation of bubbles. Therefore, the effective measures for vacuum deoxidization are as follows:

冶炼过程中碳氧反应生成的一氧化碳气泡是非自发形核,向钢液吹入惰性气体后形成很多小气泡,这些小气泡内的一氧化碳含量很少,钢液中的碳和氧能在气泡表面结合成一氧化碳进入气泡内。另外在炉底和炉壁的耐火材料表面粗糙不平,存在微小的缝隙和凹坑,也为气泡的形成创造条件。真空脱氧的有效措施包括:

(1) Before vacuum carbon deoxidization, the oxygen in steel should be in the state of easy combination with carbon, such as dissolved O_2 or Cr_2O_3, MnO and other oxides.

(1) 进行真空碳脱氧前尽可能使钢中氧处于容易与碳结合的状态,例如溶解的 O_2 或 Cr_2O_3、MnO 等氧化物。

(2) In order to accelerate the process of carbon deoxidation, the amount of argon blowing can be increased properly.

(2) 为了加速碳脱氧过程,可适当加大吹氩量。

(3) In the later stage of vacuum carbon deoxidization, aluminum and silicon are added to the molten steel to control the grain size, alloying and final deoxidization.

(3) 在真空碳脱氧的后期,向钢液中加入适量的铝和硅以控制晶粒、合金化和终脱氧。

(4) In order to reduce the oxygen content in the molten steel from refractories, the refractory with high stability should be selected in the pouring system to reduce the oxygen content in the molten steel caused by refractories entering the molten steel.

(4) 为了减少由耐火材料进入钢液中的氧量,浇注系统应选用稳定性较高的耐火材料,减少耐火材料进入钢液而造成的钢水增氧。

Task 1.5　Injection

任务 1.5　喷吹

Injection is a method of adding reactant into liquid metal with gas as carrier. The metallurgical function of the injection depends on the type of refining agent. It can complete the refining tasks such as decarburization, desulfurization, deoxidization, alloying and controlling the shape of inclusions. As a refining method, the technology of injection was first developed in France. In the early 1960s, the French Iron and Steel Research Institute reported on this aspect. With the development of refining technology outside the furnace, the application of blowing means is constantly innovated to save the consumption of alloy and control the composition of steel accurately. From the addition of bulk alloy to the addition of powder, the wire feeding technology was formed in inert atmosphere. At present, TN process in Germany and SL process in Sweden are widely used.

喷吹是用气体作载体将反应剂加入金属液内的一种手段。喷吹的冶金功能取决于精炼剂的种类,它能完成脱碳、脱硫、脱氧、合金化和控制夹杂物形态等精炼任务。作为一种精炼手段,喷吹技术最早在法国得到开发。20 世纪 60 年代初,法国钢铁研究院就有这方面的研究报道。随着炉外精炼技术的发展,喷吹手段应用不断创新,从而节省了合金料用量,精确了控制钢的成分。从块状合金的加入发展到喷粉的加入,后来形成在惰性气氛下的喂线技术。目前使用较为广泛的是德国 TN 工艺和瑞典 SL 工艺。

As shown in Figure 1-13, after entering the molten steel due to the small particle size of the refined powder, according to the principle of fluidization and gas transmission, different types of powder are injected into the molten steel or molten iron with argon (or other gases) to expand the contact area of the steel slag, so as to improve the contact area. The refining effect can be significantly improved by improving the thermodynamic and kinetic conditions of metallurgical reaction in molten steel.

喷吹法是用载气(Ar)将精炼粉剂流态化,形成气固两相流,经过喷枪,直接将精炼剂送入钢液内部。喷吹工艺示意图如图 1-13 所示。由于精炼粉剂粒度小,进入钢液后,根据流态化和气体输送的原理,用氩气(或其他气体)将不同类型的粉剂喷入钢水或铁水,从而扩大钢渣接触比面积,改善钢液内部冶金反应的热力学和动力学条件,显著提高了精炼效果。

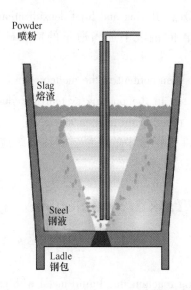

Figure 1-13 Schematic diagram of injection process
图 1-13 喷吹工艺示意图

The injection method can effectively desulfurate, deoxidize, and change the shape of inclusions, dephosphorize and alloy. The type of powder injection depends on the refining purpose. The blown gas enlarges the reaction surface area between alloy and molten steel, and solves the problem of adding active elements. And the reaction is carried out under the strong agitation of the molten pool to realize continuous feeding and control feeding. The functions realized by means of injection include:

喷吹法能有效脱硫、脱氧、改变夹杂物形态、脱磷以及合金化,喷粉的类型根据精炼目的而定。吹入的气体扩大了合金料与钢液的反应表面积,解决了活性元素的加入问题。

反应在熔池被强烈搅动下进行，实现连续供料与控制供料。喷吹手段的功能包括：

(1) The utilization of alloy elements is improved. When the powder of easily oxidizable elements (Al, Ti, B, V, Ca, Re, etc.) is injected into the molten steel as a carrier, the utilization rate of alloy elements can be increased, the burning loss can be reduced, and the composition of molten steel can be stabilized.

(1) 提高合金元素的利用率。将易氧化元素（Al、Ti、B、V、Ca、Re 等）粉剂用气作载体喷入钢液，可提高合金元素的利用率，减少烧损，稳定钢水成分。

(2) The smelting time and reduce power consumption are shortened. Especially in the process of EAF steelmaking, the oxidation period is strengthened to accelerate dephosphorization, deoxidization and desulfurization in the reduction period, the smelting time shortened and the power consumption reduced. According to the feedback of production practice, the smelting period can be shortened by 30~40 min.

(2) 缩短冶炼时间，降低电耗。尤其在电炉炼钢过程中强化氧化期加速脱磷，加速还原期脱氧脱硫，缩短冶炼时间，降低电耗。根据生产实践反馈，可缩短冶炼期 30~40min。

(3) Dephosphorization of high chromium molten steel is achieved. The dephosphorizing agent, such as CaC_2 or CaSi, can be added to the molten steel by means of blowing, which can ensure that the content of chromium, nickel, manganese and other alloy elements in the molten steel is basically unchanged, and at the same time, the dephosphorization rate of 20%~56% can be obtained to realize dephosphorization of high chromium molten steel.

(3) 实现冶炼高铬钢液的脱磷。采用喷吹的方法向钢水加入 CaC_2 或 CaSi 等脱磷剂，可以保证钢水中铬、镍、锰等合金元素含量基本不变，同时可获得 20%~56% 的脱磷率，实现高铬钢液的脱磷。

(4) The inclusion morphology is controlled. When injecting Ca-Si series powder or rare earth metal (Re) into steel water, good deoxidizing and desulfurizing effect can be achieved and inclusion morphology can be controlled at the same time.

(4) 控制夹杂物形态。向钢水中喷吹 Ca-Si 系列粉或稀土金属（Re）时，在取得良好的脱氧脱硫效果的同时，可以起到控制夹杂物形态的作用。

(5) The technological performance of molten steel has been greatly improved. After the injection of Ca-Si alloy, the technological performance of the molten steel has been greatly improved. The pouring temperature can be reduced by 10~20℃. The fluidity of the molten steel can be improved, and the phenomenon of nozzle nodulation and bottom sticking is rare.

(5) 改善钢水的工艺性能。喷吹 CaSi 合金后，钢水的工艺性能有很大改善，浇注温度可降低 10~20℃。钢水的流动性提高，很少出现水口结瘤和粘包底现象。

(6) Dusting and carburizing. An important operation in molten steel smelting is to adjust the carbon content in the steel. The coke powder is injected into the molten steel by the injection method. The carbon recovery rate is 80%~90% in the melting period and 95%~100% in the reduction period of the electric furnace. This changes the situation of increasing energy consumption by carbon addition of pig iron and reduces the cost of steel.

(6) 喷粉增碳。钢液冶炼中一项重要的操作是调整钢中碳含量，利用喷吹法向钢液喷入焦炭粉，碳的回收率在熔化期为 80%~90%，电炉还原期为 95%~100%。这样改变了由生铁增碳增加能耗的状况，并降低了钢的成本。

In production, the main ways to use the injection means are single gas injection (such as VOD process), mixed gas injection (such as AOD process), powder gas injection (such as TN process), and solid material addition (such as wire feeding process).

生产中利用喷吹手段的主要方式有单一气体喷吹（如 VOD 工艺），混合气体喷吹（如 AOD 工艺），粉气流的喷吹（如 TN 工艺），以及固体物的加入（如喂线工艺）。

Exercises

练习题

(1) What is the purpose of top slag control and what are the methods?
(1) 顶渣控制的目的是什么，有何方法？
(2) What is the refining effect of slag washing?
(2) 渣洗有何精炼作用？
(3) What are the mixing methods?
(3) 搅拌方法有哪些？
(4) What are the refining principles and functions of argon blowing?
(4) 吹氩的精炼原理是什么，吹氩精炼的作用有哪些？
(5) What are the heating methods and their characteristics?
(5) 加热方法有哪些，各有何特点？
(6) What are the main performance indexes of vacuum pump and what are the advantages of steam jet pump?
(6) 真空泵的主要性能指标有哪些，蒸汽喷射泵有哪些优点？
(7) What is spray metallurgy? And briefly describe its function.
(7) 何谓喷射冶金？并简述其作用。
(8) Describe the key points of wire feeding process are.
(8) 简述喂线工艺操作要点。
(9) How to classify inclusions in steel?
(9) 钢中夹杂物如何分类？

Project 2　Basic Process of Secondary Refining

项目2　炉外精炼基本工艺

With the development of iron and steel production technology, the quality requirements of iron and steel products are increasingly strengthened. To accurately control the content of [C], [P], [N], [H], [O] in steel, the removal of these harmful elements is limited when refining in converter or electric furnace. In order to improve the refining level, these metallurgical operations will be moved to the refining furnace. It can not only improve the quality of steel smelting and the production efficiency, but also meet the special requirements of different kinds of steel to transfer some refining tasks completed in general steelmaking furnace (converter, open hearth or electric furnace) to the 'Ladle' or special container outside the furnace.

随着钢铁生产技术的发展，钢铁产品的质量要求日益加强，要精确控制钢中的[C]、[P]、[N]、[H]、[O]含量。转炉或电炉的精炼，对这些有害元素的去除是有限的。为了提高精炼水平，这些冶金操作将移到精炼炉中进行。把一般炼钢炉（转炉、平炉或电炉）中完成的部分精炼任务，移到炉外的"钢包"或专用容器中进行。这不仅可以提高钢铁冶炼质量和生产效率，还可以满足不同钢种的特殊要求。

The project will introduce the basic process of secondary refining commonly used in iron and steel production. There are more than 30 kinds of secondary refining technologies used in production at present learned from Introduction. In order to facilitate learning, these secondary refining technologies still divided into two categories according to the development of secondary refining in the introduction: vacuum treatment technology and non vacuum treatment technology. Among vacuum treatment technology, RH series and VD series will introduced, and use vacuum technology to remove gas and inclusions in liquid steel, so as to improve the quality of liquid steel introduced later. Instead of vacuum treatment technology, ladle refining technology will be introduced with LF furnace as the core, which is developed with the goal of improving productivity and reducing product cost while ensuring quality.

本项目将介绍钢铁生产中常用的炉外精炼基本工艺。在绪论中已经了解到，目前在生产中运用的炉外精炼技术已经有30种以上，为了便于学习，还是按照绪论中按照炉外精炼技术的发展把这些炉外精炼技术分为两大类，分别为真空处理技术和非真空处理技术。其中在真空处理技术中，将介绍RH系列和VD系列：运用真空技术去除钢液气体和夹杂，以提高钢液质量。而在非真空处理技术中，将介绍以LF炉为核心的钢包精炼技术：在确保质量的同时，以提高生产率和降低产品成本为目标发展。

Ladle refining refers to the ladle refining technology and equipment, which can be used in various refining methods and has complete functions. It is usually called ladle refining furnace. It is usually called ladle refining furnace. The ladle refining furnace usually has more than three kinds of refining means, such as stirring, heating and slagging. And these means can be controlled independently. On the one hand, the purpose of refining is to improve the quality of molten steel, on the other hand, to improve the productivity and reduce the cost of furnace steel, and to transfer all the reduction refining tasks of converter and EAF to the ladle outside the furnace.

钢包精炼是指钢包精炼的技术与设备，即把多种精炼手段在钢包内完成。通常称之为钢包精炼炉。钢包精炼炉通常都具有搅拌、加热、造渣等三种以上（如真空）的精炼手段，并且这些手段都是可以独立控制。精炼的目的一方面是为了提高钢水质量，另一方面是谋求提高生产率和降低炉钢成本，将转炉、电弧炉的还原精炼任务全部移至炉外的钢包进行。

A prominent feature of ladle furnace refining is that it has heating means and can heat the molten steel in ladle. Therefore, in order to complete the heat absorption of refining task and heat loss in refining process, it can be compensated by heating, so that the ladle furnace no longer depends on the tapping temperature of primary furnace in terms of liquid steel temperature and refining time. The ladle refining furnace has the following features in terms of function:

钢包炉精炼的一个突出特点是具有加热手段，可以在钢包内对钢液进行加热。为了完成精炼任务的吸热以及在精炼过程中的散热损失，可通过加热得到补偿，使钢包炉在钢液温度和精炼时间长短方面不再依赖初炼炉的出钢温度。钢包精炼炉在功能方面具有如下特点：

(1) It has good degassing conditions. Generally, there are vacuum and stirring conditions, argon stirring and electromagnetic stirring at the bottom of ladle. The argon bubbles floating in vacuum and liquid steel provide good thermodynamic conditions for degassing.

(1) 具有良好的脱气条件。一般都具有真空和搅拌条件，钢包底吹氩气搅拌和电磁搅拌。真空和钢液内上浮的氩气泡为脱气提供良好的热力学条件。

(2) It can adjust the temperature of molten steel accurately. When heating the molten steel, the heating speed of the molten steel is controlled by adjusting the input power. In addition, in the refining process, the ladle lining fully stores heat, so that the molten steel cooling is slow in the pouring process and can be kept within the specified temperature range.

(2) 能够准确调整钢液温度。加热钢液时，通过调节输入功率来控制钢液的升温速度。此外，精炼过程中钢包内衬充分蓄热，这样在浇铸过程中钢液降温缓慢，并能保持在规定的温度范围内。

(3) The composition of liquid steel is uniform and stable. Because the agitation runs through the whole refining process, the ingredients are even and stable.

(3) 钢液的成分均匀稳定。因为搅拌贯穿整个精炼过程，所以成分均匀稳定。

(4) It can add slag to make reduction slag for refining, fully deoxidize, desulfurize and refine low sulfur steel.

(4) 可以加渣料造还原渣精炼，充分地脱氧、脱硫、精炼低硫钢种。

(5) Excellent alloying conditions. Due to the heating method, the amount of alloy added in the refining process is not limited in principle. When the alloy is added in vacuum, the yield is high and stable, and the composition of liquid steel can be controlled in a narrow range.

(5) 优越合金化条件。由于具有加热手段,原则上精炼过程中的合金加入量不受限制。在真空条件下加入易氧化的合金,其收得率高且稳定,并可以把钢液成分控制在较窄的规格范围内。

(6) Ladle refining can accomplish deoxidization, degassing, decarburization, desulfurization, purity improvement, alloying and other tasks in different degrees.

(6) 钢包精炼法在不同程度上可以完成脱氧、脱气、脱碳和脱硫,从而提高纯净度、合金化等项任务。

In the initial atmospheric ladle blowing argon, a hole is opened at the bottom of the ordinary ladle and a porous vent plug of refractory is installed. Argon blowing into the ladle through the vent plug can make inclusions in the steel float rapidly and homogenize the composition and temperature of the liquid steel. However, this simple ladle blowing argon has some shortcomings. When blowing argon, the slag on the ladle is blown away, and the molten steel is exposed to the air, resulting in the secondary oxidation of the molten steel; the unstable oxides and variable value oxides in the slag continue to interact with the molten steel; and the molten steel washes refractory materials, and constantly brings in foreign inclusions. In order to overcome the above defects of simple ladle argon blowing, ladle with immersion cover and top cover or argon blowing refining device in closed furnace are adopted.

起初的常压钢包吹氩是在普通钢包底部开一孔,并安装一个耐火材料的多孔透气塞,氩气通过透气塞吹入钢包中可促使钢中夹杂物迅速上浮,使钢液成分和温度均匀化。然而这种简单的钢包吹氩存在一些不足,吹氩时钢包上面的炉渣被吹开,钢水暴露于空气,造成钢水的二次氧化;渣中不稳定氧化物和变价氧化物继续与钢水作用;钢水冲刷耐火材料,不断带入外来夹杂物。为了克服简单钢包吹氩的缺陷,采取钢包带浸渍罩和封顶盖或在密闭炉体内吹氩精炼装置。

In 1965, Nippon Steel Co., Ltd. of Japan developed the CAB (Capped Argon Bubbling) process with ladle cover and synthetic slag, and then created the CAS (Composition Adjustment by Sealed Argon Bubbling) process for composition adjustment by sealed argon bubbling in production. On the basis of CAS process, Nippon Steel Co., Ltd. developed the CAS-OB (Composition Adjustment Seald Argon Bubbling-Oxygeon Blowing) process Blowing. In 1965, ASEA-SKF ladle refining method was developed by Sweden General Electric Company (ASEA) and Sweden bearing company (SKF). LF ladle refining method was developed by Japan special steel company in 1971. Both of them have the function of arc heating to overcome the shortcomings of cab and CAS. The main difference between them is the stirring of molten steel, the former low-frequency electromagnetic stirring is, the latter argon stirring is.

1965 年,日本新日铁公司开发出带钢包盖加合成渣吹氩精炼的 CAB 法,随后在生产中为了成分调整创造出密封吹氩的 CAS 法,在 CAS 处理法的基础上,新日铁公司开发了 CAS-OB 法。1965 年瑞典通用电气公司与瑞典轴承公司合作开发出 ASEA-SKF 钢包精炼法，日本特殊钢公司 1971 年开发出 LF 钢包精炼法，二者均具有电弧加热功能，克服了 CAB 和 CAS 法的不足。两者的主要差别在于钢水的搅拌，前者为低频电磁搅拌，后者为氩气搅拌。

Task 2.1　LF Process
任务 2.1　LF 工艺

　　LF refining is carried out by arc heating under atmospheric pressure (Ar atmosphere). The research of LF (Ladle Furnace) refining technology began in 1968. It was found that the reduction refining effect was remarkable by using the arc furnace to prepare the reduction slag, mixing the steel slag and blowing argon in the ladle. Therefore, the ladle refining technology with the function of arc heating was developed for the purpose of omitting the reduction period of the arc furnace. In 1971, the first ladle refining furnace (LF) developed by Datong Special Steel Works in Japan based on ASEA-SKF refining technology was put into operation. As shown in Figure 2-1, the method of blowing argon at the bottom of ladle is used to obtain the stirring kinetic energy of molten steel, which has the same good desulfurization and deoxidization effect as electromagnetic stirring, and is more conducive to the floating of inclusions in steel; It uses strong reducing slag for desulfurization and deoxidization, so as to realize the main metallurgical functions of inclusion control and arc heating to melt ferroalloy, adjust composition, temperature, etc.; During ladle refining, the arc is added heat, not only adjusting the temperature of molten steel, but also adding a lot of alloy. In 1973, LF was set up in BOF plant where Nippon Steel Bafan was made.

　　LF 精炼是在大气压力（Ar 气氛）下进行电弧加热。LF 精炼技术的研究始于 1968 年，当时发现用电弧炉预造还原渣、钢渣混出、钢包吹氩处理时，还原精炼效果显著，因此进行了以省略电弧炉还原期为目的的有电弧加热功能的钢包精炼技术的开发。1971 年日本大同特殊钢厂在 ASEA-SKF 精炼技术的基础上开发的第一台钢包精炼炉投入使用。1973 年，新日铁八幡制铁所在转炉厂设置了 LF。如图 2-1 所示，采用钢包底吹氩气的方法使钢液获得搅拌动能，具有与电磁搅拌同样好的脱硫、脱氧效果，更有利于钢中夹杂物上浮；它用强还原性渣脱硫、脱氧，进而实现夹杂物控制和电弧加热熔化铁合金，调整成分、温度等主要冶金功能；钢包精炼时进行电弧加热，不仅可以调整钢水的温度，而且可以加入大量的合金。

　　LF furnace does not have a vacuum system, but because the ladle and furnace cover are sealed and air is isolated, the graphite electrode reacts with FeO, MnO and Cr_2O_3 in the slag to generate CO gas during heating, which reduces the oxygen content in the atmosphere of LF furnace. In the refining process, diffusion deoxidation and precipitation deoxidation are used to create reduction conditions of molten steel, which can further deoxidize, desulfurize and remove non-metallic inclusions. In addition, argon stirring can accelerate the material transfer between steel and slag, which is beneficial to the deoxidization and desulfurization of molten steel. For example, argon blowing can accelerate the floating speed of Al_2O_3 inclusions. In sealed LF furnace, after

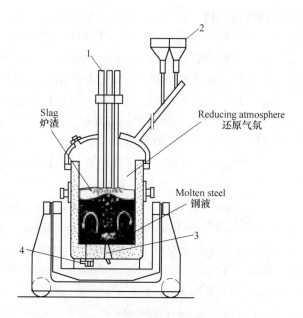

Figure 2-1 LF ladle refining furnace
图 2-1 LF 钢包精炼炉

1—Electrode; 2—Alloy hopper; 3—Permeable brick; 4—Sliding nozzle
1—电极; 2—合金料斗; 3—透气砖; 4—滑动水口

15min of argon blowing, Al_2O_3 inclusions larger than 20μm in steel can be basically removed.

LF 炉本身不具备真空系统，但由于钢包与炉盖密封，隔离空气，加热时石墨电极与渣中 FeO、MnO、Cr_2O_3 等反应生成 CO 气体，使 LF 炉内气氛中氧含量减少。精炼过程通过扩散脱氧和沉淀脱氧造成钢液的还原条件，可以进一步脱氧、脱硫及去除非金属夹杂。另外，氩气搅拌加速钢、渣之间物质传递，有利于钢液脱氧、脱硫反应。例如，吹氩可以加速 Al_2O_3 夹杂物上浮速度，在密封的 LF 炉，吹氩 15min 后，可使钢中大于 20μm 的 Al_2O_3 夹杂基本清除。

The development of EAF slag free tapping and converter slag retaining tapping technology has greatly promoted the development and improvement of LF refining technology. However, it is quite difficult to realize slag free tapping of EAF and less slag tapping of converter in actual production. Therefore, the slag modification processes after tapping commonly used in current EAF process and converter process appear: EAF or converter tapping→steel slag modification (adding aluminum, slag, Ca-Si or slag modifier)→LF refining. The first stage of EAF development is the traditional EAF including melting, oxidation and reduction. The second stage is due to the type of EAF (tapping trough EAF). In order to prevent the molten steel from being polluted by oxide slag and to play the role of deoxidizing and desulfurizing, the reduction slag must be made in the EAF, mixed out and further reduction refining can be completed by LF. The third stage is due to the development of slag free tapping technology. The reduction period is all completed by LF refining, that is to say, the form of EAF—LF—CC is formed.

电炉无渣出钢和转炉挡渣出钢技术的发展，为 LF 精炼技术的发展与完善起到巨大的推动作用。但是在实际生产中要实现电弧炉的无渣出钢和转炉的少渣出钢相当困难，因此出现了目前在电炉流程和转炉流程中普遍采用的出钢后对下渣改性处理工艺，即：电弧炉或转炉出钢→钢内渣改性（加铝、加渣料、加 Ca-Si 或加改渣剂）→LF 精炼。电弧炉发展的第一阶段包括熔化、氧化、还原的传统型电弧炉。第二阶段由于电弧炉炉型（出钢槽式电弧炉）的原因，为避免氧化渣污染钢液及发挥钢渣脱氧、脱硫的作用，电弧炉内必须造好还原渣，钢渣混出，由 LF 完成进一步还原精炼的任务。第三阶段由于无渣出钢技术的开发，还原期全部由 LF 精炼来完成，即形成了现代电弧炉炼钢流程电炉—LF—连续铸钢的形式。

As the EAF-LF-CC mode greatly improves the productivity of EAF, and the adoption of near net shape continuous casting, the EAF steelmaking method of producing common steel with high efficiency and short process has also been greatly developed. LF refining can produce special steel (alloy steel) in converter plant, and the large-scale production system of multiple varieties and high-quality steel composed of 'Rough Refining-Secondary Refining' has been determined. With strong refining function, it is suitable for the production of ultra-low sulfur and ultra-low oxygen steel, with arc heating function, high thermal efficiency, large range of temperature rise, and accurate control of high temperature. In addition, with stirring and alloying function, it is easy to achieve narrow composition control and improve the stability of products; With slag steel refining process, the refining cost is low; The equipment is simple, and the investment is less.

由于电炉—LF—连续铸钢方式使电炉生产率大大提高。近终形连铸的采用，扩大了高效短流程生产普通钢的电炉炼钢法的发展。在转炉工厂采用 LF 精炼，能够生产特殊钢（合金钢），确定"粗精炼—炉外精炼"构成的多品种、高质量钢的大生产体制。电炉—LF—连续铸钢方式精炼功能强，适宜生产超低硫、超低氧钢；具备电弧加热功能，热效率高，升温幅度大，温度控制精度高，控温准确度可达±5K；具备搅拌和合金化功能，易于实现窄成分控制，提高产品的稳定性；采用渣钢精炼工艺，精炼成本较低；设备简单，投资较少。

2.1.1 LF Production Process

2.1.1 LF 生产工艺过程

LF furnace has become one of the main refining methods for pure steel in China because of its advantages of complete metallurgical functions, simple structure, convenient operation and less investment. It plays a leading role in the secondary refining and greatly improves the secondary refining ratio in China. LF refining can improve the purity of molten steel and meet the requirements of continuous casting on the composition and temperature of molten steel, and become the main matching equipment between steelmaking and continuous casting. LF method can be matched with converter or arc. The main production processes include EBT tapping of converter and electric furnace, adding alloy and slag (2% lime, fluorite, etc.) during tapping, bottom blowing argon, electrifying and heating up, slagging, sampling and analysis in 10min, adding slag (1%), temperature measurement and sampling, adding alloy to see deoxidation, and preparing for tapping. Smelting time is generally 30~50min and power consumption is 50~80kW·h/t, which is the link between modern converter, electric furnace and continuous casting, as shown in Figure 2-2.

由于其冶金功能齐全,结构简单,操作方便,投资少等优点,LF 炉已经成为我国纯净钢的主要炉外精炼方法之一,在炉外精炼中占主导地位,大大提高了我国的炉外精炼比。LF 精炼可提高钢液的纯净度,满足连铸对钢液成分及温度的要求,成为炼钢与连铸间匹配的主要设备。LF 法可以与转炉或电弧匹配,其主要生产过程为:转炉、电炉 EBT 出钢,出钢过程加合金、加渣料(石灰、萤石等 2%),底吹氩、通电升温、化渣,10min 取样分析,加渣料(1%),测温取样,加合金看脱氧,准备出钢。冶炼时间一般为 30~50min,电耗 50~80kW·h/t,是现代转炉、电炉与连铸联系的纽带。LF 精炼生产流程如图 2-2 所示。

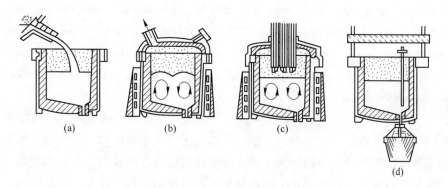

Figure 2-2 LF refining production process
图 2-2 LF 精炼生产流程
(a) Holding molten steel; (b) Degassing and stirring; (c) Arc heating and stirring; (d) Casting
(a) 盛接钢水;(b) 脱气搅拌;(c) 电弧加热搅拌;(d) 浇铸

2.1.1.1 Argon Blowing in Ladle

2.1.1.1 钢包吹氩

The ladle argon blowing starts from tapping until the ladle is hoisted to LF waiting station. The metallurgical purposes of argon stirring at this stage include:

钢包吹氩从出钢开始,一直到钢包吊往 LF 等待工位。此阶段吹氩搅拌的冶金目的包括:

(1) Promote the melting and dissolving of the alloy and slagging agent added in tapping.

(1) 促进出钢加入的合金与造渣剂的熔化溶。

(2) Uniform bath temperature.

(2) 均匀熔池温度。

(3) The deoxidation products in tapping process are removed.

(3) 去除出钢过程的脱氧产物。

(4) Strengthen the mixing of slag and steel to reduce the content of S in the molten steel.

(4) 加强渣、钢混合,降低钢液中的 S 含量。

2.1.1.2　Ladle to LF Waiting Station

2.1.1.2　钢包到 LF 等待工位

After the ladle arrives at the LF waiting position, connect the argon blowing pipe. At this time, ensure the proper argon blowing amount to avoid the exposed steel level, and at the same time, ensure that no slag splashes out of the ladle. If the amount of tapping is too large or there is a lot of slag, a part of liquid steel or slag should be poured out. If the slag surface can't be blown, the pressure shall be increased immediately to blow argon or the accident argon gun shall be used to blow argon and open the porous brick. If it can't be blown open, it needs to be rewound. For the production of high-quality steel deoxidized by aluminum, it is better to feed aluminum at the waiting station, and change the dissolved oxygen in the molten steel into oxide inclusions as soon as possible, so as to provide a longer time for the removal of inclusions and reduce the total oxygen content in the molten steel.

钢包到 LF 等待工位后,接通吹氩管,这时吹氩要保证合适的吹氩量,以避免钢液面裸露,同时保证不要把钢渣溅出钢包。如果出钢量过大或下渣较多,应倒出一部分钢液或下渣。如果渣面吹不开,就要瞬间增大压力吹氩或用事故氩枪吹氩,吹开多孔砖。如果还吹不开,就要进行倒包处理。对于生产铝脱氧的高质量钢,最好在等待工位喂铝,尽早把钢液中的溶解氧全部变成氧化物夹杂,为夹杂物的去除提供较长的时间,从而降低钢液中的全氧量。

2.1.1.3　Slagging

2.1.1.3　造渣

When the ladle is at the heating position, and charging from the silo, the argon flow will be increased, the slag surface blown, and the slag forming material added to the exposed steel surface. When the slag is heated at the same time, lime and Al_2O_3 are added, or even CaC_2, SiC or aluminum particles are added to deoxidize the slag. After heating for 3~5min, the slag is observed through the slag gate to ensure the good fluidity of the slag. If the slag is too thick, add more bauxite or fluorite; and if the slag is too thin, add lime. Take the slag sample to judge the deoxidation condition (When solidified, it is gray white, indicating that the slag deoxidizes well; and when the slag blackens, add aluminum particles or other deoxidizers to continue to reduce the oxygen in the slag.)

钢包到加热位置,当要从料仓加料时,应增加氩气流量,吹开渣面,把造渣料加到裸露的钢液面上。在加热的同时处理渣,加入石灰和 Al_2O_3,甚至加入 CaC_2、SiC 或铝粒进行渣脱氧,加热 3~5min 后,通过渣门观察渣的情况,保证渣的流动性。若渣太稠,则多加铝矾土或萤石；若渣太稀，则加入石灰。取渣样判断脱氧情况（凝固时呈灰白色，表示渣脱氧良好；渣发黑，加铝粒或其他脱氧剂继续降低渣中的氧）。

2.1.1.4 Ladle Sampling

2.1.1.4 钢包取样

After the slag is heated and treated (the slag is basically white), the temperature measurement is the same as that of the first.

加热及处理渣后（渣基本变白），测温取第一样。

2.1.1.5 Add Alloy, Homogenize and Adjust Temperature

2.1.1.5 加入合金，均匀化及调整温度

According to the analysis results of the amount of alloy added in tapping and ladle, the amount of alloy added is determined to reach the required composition of the finished steel. The added alloy will change the composition of molten steel according to the predetermined alloy yield. If the composition of molten steel fails to change according to the amount of added alloy, it indicates that the deoxidization of molten steel is incomplete, and deoxidizer such as aluminum wire is needed to ensure the optimal yield of alloy elements. After adding the alloy, continue heating and stirring for 5min to ensure that the added alloy dissolves. If the expected liquid steel composition is not obtained, a new alloy must be added to meet the composition requirements of the steel grade. Aluminum decreases with the refining process. If aluminum decreases rapidly, it indicates that aluminum oxidation rate is very high. At this time, aluminum should be fed to reduce the dissolved oxygen in steel and ensure the stability of aluminum in molten steel. Desulfuration is an important function of ladle furnace process. The best condition for desulfuration is to ensure high basicity and bath temperature higher than 1580℃. For the production of sulfur-containing steel, due to the deep deoxidization and slagging in the refining process, the sulfur content has been reduced to a lower level, so it is necessary to control the sulfur content in the steel by feeding sulfur wire at the end of ladle treatment.

根据出钢加入的合金量及钢包第一样分析结果，确定加入的合金量是否达到成品钢要求的成分。加入的合金应按预定的合金收得率改变钢液成分，如果钢液成分未能按加入的合金数量改变，说明钢液脱氧不完全，则需要用铝线等脱氧剂脱氧以确保合金元素的最佳收得率。加入合金后，继续加热并搅拌5min，以确保加入的合金溶解，如果没有得到预期的钢液成分，则必须加入新合金，以满足钢种的成分要求。铝随着精炼过程的进行而减少，如果铝迅速降低，则表明铝氧化速率很高，这时应补喂铝，降低钢中的溶解氧，从而保证钢液中铝的稳定性。脱硫是钢包炉工艺的重要功能，脱硫的最佳条件是保证高碱度及熔池温度高于1580℃。对于生产含硫的钢种，由于精炼过程中深脱氧和造渣，硫含量降到了较低的水平，所以必须在钢包处理末期通过喂硫线控制钢中硫含量。

2.1.1.6　Modification of Inclusions

2.1.1.6　夹杂物变性处理

After the last feeding, mix with argon for 3~6min. If the stirring time is too short, the composition and temperature will not be uniform; and if the stirring time is too long, the secondary oxidation of the molten pool will occur. After the composition of the molten steel is qualified, the silicon calcium wire or iron calcium wire is fed for inclusion denaturation. The cleanliness of steel depends on how deoxidizing products and other inclusions are absorbed by slag, while castability depends on the calcium treatment of inclusions not absorbed by slag. In order to obtain high cleanliness and good pouring performance, the best condition is that the molten steel is fed with appropriate silicon calcium wire or iron calcium wire and stirred by argon blowing under the condition of good composition and fluidity without breaking the slag layer. After the alloy is added for the last time, it shall be stirred with argon for 5min, and then fed with wire.

最后一次加料后，吹氩搅拌 3~6min。若吹氩搅拌时间太短则不能均匀成分与温度；若太长则会产生熔池的二次氧化。钢液成分合格后，喂入硅钙线或铁钙线进行夹杂物变性操作。钢的洁净度取决于脱氧产物及其他夹杂物如何被渣吸收，而可浇注性则取决于未被渣吸收的夹杂物的钙处理。要获得高洁净度及良好的浇注性能，最佳条件是钢液在符合成分、流动性好并不打破渣层的情况下，喂入合适的硅钙线或铁钙线吹氩搅拌。在最后一次加入合金后，应吹氩搅拌 5min，再喂线。

2.1.1.7　Stop Argon and Lift the Ladle to Continuous Casting

2.1.1.7　停氩，钢包吊往连铸

The stirring time of weak argon blowing will be more than 5min after feeding calcium silicon wire or calcium iron wire. Measure the temperature and take samples to analyze the final composition of the molten steel. After that, the ladle car leaves the heating station, adds ladle heating agent or covering agent (If VD treatment is carried out, no ladle heating agent or covering agent is added), takes off the argon blowing pipe, stops argon blowing, and lifts it to the continuous casting table for pouring or VD treatment.

喂入硅钙线或铁钙线后弱吹氩搅拌时间应大于 5min，测温，取样，分析最终钢液成分。之后钢包车开出加热工位，加入钢包发热剂或覆盖剂（如果进行 VD 处理，不加钢包发热剂或覆盖剂），取掉吹氩管停止吹氩，并吊往连铸台浇注或进行 VD 处理。

At present, many converter plants only play the role of composition and temperature adjustment, they often unable to control the cleanliness of liquid steel, especially in small and medium-sized converter plants. On the one hand, it is difficult to control the amount of slag and there is no effective slag changing process; On the other hand, converter smelting time is fast and refining time is short. These two points make it difficult for medium and small converter with LF refining to produce high clean steel.

目前很多转炉厂LF只起成分及温度调整的作用,往往达不到对钢液洁净度的控制,特别是中小转炉厂。一方面是由于下渣量控制困难,以及没有进行有效变渣处理工艺;另一方面是转炉冶炼时间节奏快,精炼时间短。这两点使得中小转炉配LF精炼生产高洁净钢较为困难。

2.1.2　The Main Process System

2.1.2　主要工艺制度

The main operations of LF refining include: The feeding operation of adding aluminum amount is determined according to the requirements of acid soluble aluminum in steel and the content of dissolved oxygen in molten steel. The slag making system of submerged arc heating, desulfurization and absorption of inclusions is considered; The argon blowing and stirring treatment to prevent aspiration, slag rolling and accelerate the removal of inclusions; The arc heating is considered controlled by temperature target; and consider the composition of molten steel with the lowest cost fine adjustment. So the process system needs to be established:

LF精炼的主要操作包括:根据钢中酸溶铝的要求及钢液中的溶解氧含量,确定加铝量的喂铝线操作;考虑埋弧加热、脱硫、吸附夹杂物的造渣制度;考虑防止吸气、卷渣以及加快夹杂物去除的吹氩搅拌处理;考虑温度目标控制的电弧加热;考虑最低成本的钢液成分微调。LF精炼需要建立以下工艺制度:

(1) Heating and temperature control: Three electrodes are inserted into the slag layer for submerged arc heating in LF furnace. This method has the advantages of small radiant heat, high thermal efficiency, general heating efficiency of 60%, and higher than the heating efficiency of electric furnace. The power consumption of each ton of steel is $0.5 \sim 0.8 kW \cdot h$ when it is heated up to 1℃. The temperature speed is determined by the specific power of power supply ($kV \cdot A/t$), and the specific power of power supply is determined by the melting loss index of ladle refractory. Usually, the power supply of LF furnace is $150 \sim 200 kV \cdot A/t$, and the heating rate can reach $3 \sim 5$℃/min. The submerged arc foam technology can increase the heating efficiency by 10%~15%. The computer dynamic control of the terminal temperature can ensure the control accuracy, which is less than 5℃. When heating, graphite electrode is inserted into the slag layer, and the arc is generated in the slag layer. Using the submerged arc heating method, the shielding effect of the slag on the arc reduces the thermal radiation of the arc to the lining, and has protective effect on the lining. At the same time, molten steel and slag effectively absorb arc heat and improve thermal efficiency.

(1) 加热与温度控制。LF炉采用三根电极插入渣层中进行埋弧加热。这种方法辐射热小,对炉衬有保护作用,热效率高,加热效率一般≥60%,高于电炉升温热效率。吨钢水平均升温1℃耗电$0.5 \sim 0.8 kW \cdot h$。升温速度决定于供电比功率($kV \cdot A/t$),供电比功率的大小又决定于钢包耐材的熔损指数。通常LF炉的供电比功率为$150 \sim 200 kV \cdot A/t$,升温速度可达3~5℃/min。采用埋弧泡沫技术可提高加热效率10%~15%,采用计算机动态控制终点温度可保证控制精度不大于±5℃。加热时将石墨电极插入泡沫渣层中,电弧在渣层中产生。采用埋弧加热法,泡沫渣对电弧的屏蔽作用减少了电弧光对包衬的热辐射,对包衬有保护作用,与此同时钢液和炉渣有效地吸收电弧热,从而提高了热效率。

In order to ensure the temperature stability of continuous casting molten steel and the matching of superheat and drawing speed, it is necessary to minimize and stabilize the temperature drop in the refining process, and play the functions of furnace temperature rise, constant temperature and buffer. In actual production, the allowable fluctuation of tapping temperature of primary furnace and the temperature of molten steel leaving the refining station is ±10℃. Because of the matching problem of furnace and machine and the large temperature drop of tundish, the temperature of the first and second furnaces is higher in tapping and outgoing. The factors influencing the temperature of molten steel in LF refining are as follows:

为保证连铸钢水温度稳定,过热度与拉速匹配,则需要最大限度减少和稳定精炼过程的温降,发挥精炼炉升温、恒温和缓冲的功能。在实际生产中,初炼炉出钢温度和钢水离开炉精炼站温度的允许波动为10℃。开浇第一、二炉时,由于炉机匹配问题和中包存在较大的温降,其在出钢和出站温度还要更高一些。影响LF炉外精炼钢液温度的因素包括:

1) LF furnace heat loss, include: heat dissipation on the surface of ladle, heat storage of ladle refractory, heat absorbed by argon used for argon blowing and stirring, heat taken away by flue gas and smoke, and heat taken away by cooling water of furnace cover. The heat budget of LF furnace mainly refers to arc heating, reaction heat of slag steel and slag forming heat.

1) LF炉热损失,包括:包体表面散热,钢包耐火材料蓄热,吹氩搅拌所用氩气吸收的热量,烟气、烟尘带走的热量,炉盖冷却水带走的热量。LF炉热收支项主要指电弧加热,渣钢反应热和成渣热。

2) Heat change caused by alloy addition: The effect of the alloy added into the molten steel on the temperature of the molten steel mainly depends on the amount of alloy added. A large amount of alloy is added at a time will significantly reduce the temperature of the molten steel. The alloy is heated by the molten steel after being added into the molten steel, rising from the initial ambient temperature to the same temperature as the molten steel; and then it absorbs the heat of the molten steel and melts, and dissolves in the molten steel. The effect of adding 1% alloy element into the molten steel on the temperature of the molten steel.

2) 合金加入引起的热量变化:加入钢液内的合金对钢液温度的影响主要取决于合金加入量,一次加入大量合金将显著降低钢液温度。合金加入钢液后被钢液加热,从初始环境温度上升到与钢液相同温度;然后吸收钢液的热量熔化,并溶解于钢液内。钢液中加入1%合金元素时对钢液温度的影响。

In actual production, due to the influence of slag quantity, ladle condition, fan opening and other conditions, the calculation results are different from the actual temperature. According to the heat balance, the temperature of molten steel in the refining process can be calculated. And the empirical formula for temperature control of molten steel can be obtained through a large number of data statistical regression analysis.

在实际生产中,由于受到渣料加入数量、钢包状况和风机开度等条件的影响,计算结果与实际温度有一定的出入。根据热量平衡,可以计算出精炼过程中钢液温度,也可以通过大量数据统计回归分析,得到钢液温度控制经验公式。

For example, the empirical formula of temperature within 15min after heating in a 70t LF furnace of a factory are given by

例如,某厂 70t LF 炉加热 15min 内的温度经验公式为

$$T = 25 - 2D + 1.8t \tag{2-1}$$

$$T = (10 - D)t \tag{2-2}$$

Where　D——the temperature rise gear in ℃;

　　　　t——the temperature rise time in min.

式中　D——升温档位,℃;

　　　t——升温时间,min。

3) The effect of ladle preheating temperature on temperature drop of molten steel: In order to reduce the temperature drop of molten steel in the process of tapping, it is necessary to preheat the ladle before tapping. The preheating temperature of ladle has a great influence on the temperature drop of molten steel. The higher the preheating temperature of ladle wall, the smaller the temperature drop of molten steel. For the ladle with preheating temperature of 500℃ and 900℃, the difference of temperature drop of molten steel is about 50℃. In the first 20min, the temperature of molten steel almost decreased in a straight line, and after 35min, the wall heat storage reached saturation. The temperature of molten steel is still lower than that of unheated steel after it is heated with low power for a period of time in LF furnace, because the heat stored in the lining is greater than that supplied by the electrode.

3) 钢包预热温度对钢液温降的影响:为了降低出钢过程中钢液的温降,出钢前需要对钢包进行预热,钢包预热温度对钢液温降的影响较大,包壁预热温度越高,钢液的温降越小。预热温度为 500℃与预热温度为 900℃的钢包,钢液温降相差约 50℃。前 20min 内钢液温度几乎呈直线下降,35min 后包壁蓄热基本达饱和。钢液在 LF 炉小功率加热一段时间后,钢液温度仍比未加热时低,这是因为包衬蓄热量大于电极供给热。

4) The effect of slag thickness on the temperature drop of molten steel: In order to reduce the heat loss of molten steel, a certain thickness of slag should be covered on the surface of molten steel. The thickness of slag layer has great influence on the heat dissipation of slag surface. The thicker the slag layer is, the less heat the surface of molten steel is lost from; The thinner the slag layer is, the greater the heat dissipation of the surface. When the slag thickness is more than 50mm, the heat loss of slag surface is basically the same for different slag layer thickness. Therefore, in order to reduce the heat loss of molten steel, it is necessary to ensure the slag layer thickness greater than 50mm. The surface temperature of slag drops rapidly, and then drops below 900℃ in 5min. After 20min, the temperature drop gradually decreased, even did not change, that is, the surface temperature of slag remained unchanged, indicating that the heat dissipation of slag reached a stable state, the heat loss through the slag surface was less, and the slag prevented the heat loss of molten steel.

4) 渣层厚度对钢液温降的影响:为了减少钢液的散热损失,钢液面上需要覆盖一定厚度的渣子。渣层厚度对渣表面散热量影响较大。渣层越厚,由钢液表面散失的热量越

少；渣层越薄，表面散热量越大。当渣厚大于50mm时，不同渣层厚度对渣表面的热损失基本相同。所以从减少钢液热损失方面考虑，有必要保证大于50mm的渣层厚度。渣表面温度降得很快，5min内便降到900℃以下；20min以后温降逐渐减小，甚至没有变化。渣表面温度基本保持不变，表明此时渣散热达稳定状态，通过渣表面损失的热量较少，渣阻止了钢液的热量损失。

5) The effect of argon blowing and stirring on temperature drop of molten steel: Because the temperature gradient of the molten steel in the ladle is eliminated by the argon blowing stirring, the temperature of the molten steel in the upper, middle and lower layers of the ladle is more uniform, and the heat transfer ability of the molten steel to the ladle wall is increased; When the argon blowing stirring is too intense, the liquid surface of the steel breaks through the slag layer with exposed, and the heat dissipation is increased. The main reason for the temperature drop of molten steel is the increase of heat dissipation caused by argon blowing and stirring.

5) 吹氩搅拌对钢液温降的影响：由于吹氩搅拌消除了包内钢液的温度梯度，使钢包上、中、下层的钢液温度更加均匀，从而增加了钢液向包壁的热传导能力；在吹氩搅拌过分激烈时，钢液面冲破渣层而裸露，散热量增加。吹氩搅拌使散热量增加是引起钢液温降的主要原因。

(2) White slag refining process: In the operation of LF furnace, the white slag is formed by strengthening deoxidation of slag. The deoxidation effect of molten steel is achieved due to the adsorption and dissolution of oxide in molten steel by slag. In addition, due to temperature compensation and strong stirring of argon blowing, LF furnace can produce low sulfur steel or ultra-low sulfur steel with the increase of basicity and sulfur distribution ratio in slag.

(2) 白渣精炼工艺：LF炉操作中通过对炉渣强化脱氧形成白渣，由于渣对钢液中氧化物的吸附和溶解，使得钢液脱氧效果好。另外，由于有温度补偿，吹氩强烈搅拌，随渣中碱度提高，S的分配比增大，LF炉可精炼出低硫钢或超低硫钢。

The white slag is used for refining to realize desulfurization, deoxidization and production of ultra-low sulfur and low oxygen steel. White slag refining is the core of LF furnace process operation: Tapping to block slag and the amount of slag is the control of below 5kg/t; The ladle slag upgrading to control is $R \geqslant 2.5$ and $w(TFe+MnO) \leqslant 3.0\%$; Generally, Al_2O_3-CaO-SiO$_2$ system slag is used for white slag refining to control $R \geqslant 4$ and $w(TFe+MnO) \leqslant 1.0\%$. The atmosphere in furnace is controlled to be weak oxidation to avoid re oxidation of slag, and proper mixing to avoid steel surface exposure and ensure melting. There is a high mass transfer rate in the cell.

利用白渣进行精炼，实现脱硫、脱氧、生产超低硫和低氧钢。白渣精炼是LF炉工艺操作的核心：出钢挡渣，控制下渣量≤5kg/t；钢包渣改质，控制$R \geqslant 2.5$，渣中$w(TFe+MnO) \leqslant 3.0\%$；白渣精炼，一般采用$Al_2O_3$-CaO-SiO$_2$系炉渣，控制$R \geqslant 4$，渣中$w(TFe+MnO) \leqslant 1.0\%$。控制炉内气氛为弱氧化性，避免炉渣再氧化；适当搅拌，避免钢液面裸露，并保证熔池内具有较高的传质速度。

(3) Alloy fine tuning in composition control: The composition of molten steel is adjusted by adding alloy material under arc heating, bottom blowing argon stirring and reducing atmosphere. The alloy yield is high, and the alloy element can be quickly uniform in the molten steel after

melting. The alloy composition can be adjusted to a narrow range by adding alloy material and analyzing at the same time.

（3）合金微调于成分控制：钢液成分的调整是在电弧加热、底吹氩搅拌和还原气氛下加入合金料，合金收得率高，合金元素熔化后能很快在钢液内均匀。通过边加合金料边分析，微调合金成分，把合金成分调整到很窄的范围内。

On line establishment of rapid analysis facilities to ensure the analysis of the corresponding time, which is no more than 3min; accurate estimation of molten steel weight and alloy yield, good deoxidation of molten steel, white slag refining, and computer online accurate calculation of various alloy addition to ensure the accuracy and stability of molten steel composition.

在线建立快速分析设施，保证分析相应时间不多于3min；精确估算钢水质量和合金收得率，钢水脱氧良好、实现白渣精炼，以及用计算机在线准确计算各种合金加入量，从而保证钢水成分的准确性与稳定性。

（4）Refining of high basicity synthetic slag: In LF furnace, high basicity synthetic slag is produced in reducing atmosphere for refining. Through the refining effect of synthetic slag, the content of oxygen, sulfur and inclusions in steel can be reduced. It can reduce [O] to 0.002%, [S] to 0.005% or even 0.003%. The synthetic slag has strong reducibility in LF furnace, which is the result of the interaction between the good reductive atmosphere and argon stirring in LF furnace. Generally, the slag content is 2%~8% (by mass) of the liquid steel content. When smelting in LF furnace, deoxidizer can not be added, but the purpose of deoxidization can be achieved by the absorption of oxide by synthetic slag.

（4）高碱度合成渣精炼：LF炉利用还原气氛下造高碱度合成渣进行精炼，通过合成渣的精炼作用可以降低钢中氧、硫及夹杂物含量（质量分数），把［O］降至0.002%，［S］降至0.005%甚至0.003%。合成渣在LF炉内具有很强的还原性，这是由于LF炉内具有良好的还原气氛和氩气搅拌，互相作用的结果。一般渣量为钢液量的2%~8%。LF炉冶炼时可以不用加脱氧剂，而是靠合成渣对氧化物吸附，从而达到脱氧的目的。

However, it should be noted that the refining slag is usually divided into high basicity slag and low basicity slag. Generally, the slag with a basicity (CaO/SiO_2) greater than 2 is high basicity slag. The high basicity slag is suitable for the secondary refining of general aluminum killed steel and has good effect in the desulfurization of molten steel. For steel with special requirements, such as cord steel, steel wire rope steel, bearing steel, etc., low alkalinity slag will be used. In order to avoid the formation of too many alumina inclusions in the deoxidization process, Si-Mn deoxidization and neutral refining slag are used in most of these steels. After refining, circular or elliptical composite inclusions with low melting point are formed, which can deform during processing and cause less harm.

然而，通常精炼渣分为高碱度渣和低碱度渣，一般碱度（CaO/SiO_2）大于2为高碱度渣，高碱度渣适用于一般铝镇静钢二次精炼，在钢水脱硫等方面具有较好的效果。对于具有特殊要求的钢种，如帘线钢、钢丝绳钢、轴承钢等，需采用低碱度渣。为了避免在脱氧过程中生成过多氧化铝夹杂，大多采用Si-Mn脱氧，采用中性精炼渣，精炼后形成较低熔点的圆形或椭圆形复合夹杂物，在加工时可以变形，危害较小。

(5) Argon stirring process: From the ladle entering LF station, it is necessary to carry out the whole process of argon blowing operation; and in the smelting process, different argon flow rates should be selected, especially in the middle stage of smelting, to create the dynamic conditions for deep desulfurization. And argon flow control is particularly important in production to prevent the liquid steel from carbon and nitrogen absorption.

(5) 氩搅拌工艺：从钢包进入 LF 站开始，就要进行全程吹氩操作；在冶炼过程中，要选择不同的氩气流量，尤其是在冶炼中期，既要创造深脱硫的动力学条件，又要防止钢液增碳及吸氮。氩气流量控制在生产中尤为重要。

In the process of deep desulfurization, with the decrease of sulfur content in the steel, the reaction speed will also decrease. At this time, it is necessary to extend the treatment time. The treatment time of LF should be controlled according to the production rhythm and the degree of deep desulfurization. In the actual production, the treatment time of LF furnace is 40~50min. The trace elements (such as niobium, vanadium and titanium) in steel are selected to be added in the later stage of refining, and sufficient weak mixing time is ensured.

在进行深脱硫时，随钢中硫含量的降低，反应速度也随之降低。此时延长处理时间非常必要，要根据生产节奏和深脱硫的程度来控制 LF 的处理时间。实际生产中控制 LF 炉的处理时间为 40~50min，钢中微量元素（如铌、钒、钛）在精炼后期加入，并保证充分的弱搅拌时间。

Argon stirring is good for the chemical reaction between steel and slag, it can accelerate the material transfer between steel and slag, and it is also good for the deoxidation and desulfurization of liquid steel. Argon blowing can also remove non-metallic inclusions, especially for the removal of Al_2O_3 type inclusions. Argon blowing can accelerate the floating speed of Al_2O_3 particles. In particular, it is worth mentioning that the stirring of argon blowing in LF furnace is carried out in a sealed furnace excluding atmosphere, so the amount of argon blowing can also be increased. Starting from the ladle entering LF station, the whole process of argon blowing will be carried out. In the smelting process, different argon flow rates will be selected. Especially in the middle stage of smelting, the dynamic conditions of deep desulfurization will be created, and the carbon and nitrogen absorption of molten steel will be prevented. The argon flow control is particularly important in production: In the operation process of ladle refining furnace, it will be based on each stage in order to give full play to the function of LF, and the air volume of bottom blowing is adjusted according to different characteristics. During slagging, large flow rate, uniform composition and temperature do not need large stirring work and blowing flow, but in desulfurization stage, the air volume of bottom stirring can be properly increased, and the slag droplets can be drawn into the steel water to form a large slag steel contact area and accelerate the reaction. In heating stage, large stirring power should not be used, so as to avoid the instability of the arc and the large amount of stirring power and temperature dropped. For the removal of non-metallic inclusions in the steel at the later stage of treatment, the bottom blowing rate should be reduced to prevent the slag from causing new pollution to the molten steel. In the process of deep desulfurization, with the decrease of sulfur content in the steel, the reaction speed will also decrease, so it is necessary to extend the treatment time. The treatment time of LF should be controlled according to the production rhythm and the degree of deep desulfurization.

氩气搅拌有利于钢、渣之间的化学反应。它可以加速钢、渣之间的物质传递,有利于钢液的脱氧,脱硫反应的进行。吹氩搅拌还可以去除非金属夹杂物,特别是对 Al_2O_3 类型的夹杂物上浮的去除更为有利,吹氩可加速 Al_2O_3 粒子的上浮速度。特别值得提出的是 LF 炉的吹氩搅拌是在排除大气的密封炉内进行,因此还可以加大吹氩量。从钢包进入 LF 站开始,就要进行全程吹氩操作,并且在冶炼过程中选择不同的氩气流量,尤其是在冶炼中期,既要创造深脱硫的动力学条件,又要防止钢液的增碳和吸氮。氩气流量的控制在生产中尤其重要。在钢包精炼炉的操作过程中,应该根据各阶段不同的特点调整底吹搅拌气量,以充分发挥 LF 的功能。化渣期间采用大流量,之后均匀成分和温度阶段搅拌功和吹气流量降低。但在脱硫阶段,则可以适当提高底吹搅拌的气量,将炉渣液滴卷入钢水中形成较大的渣钢接触面积,加速反应的进行。在加热阶段中不应使用大的搅拌功率,以免引起电弧的不稳定,从而造成大的温降。对于处理后期的去除钢中非金属夹杂物,则应该减小底吹气量,防止卷渣造成对钢水新的污染。在进行深脱硫时,随钢中硫含量的降低,反应速度也随之降低,因此延长处理时间非常必要,要根据生产节奏和深脱硫的程度控制 LF 的处理时间。

In the actual production, the treatment time of LF furnace is 40~50min. The trace elements (such as niobium, vanadium and titanium) in steel are selected to be added in the later stage of refining, and sufficient weak mixing time is ensured.

实际生产中控制 LF 炉的处理时间为 40~50min,钢中微量元素(如铌、钒、钛)在精炼后期加入,并保证充分的弱搅拌时间。

Another function of argon stirring is to accelerate the temperature and composition uniformity in the molten steel, and to accurately adjust the complex chemical composition, which is essential for high-quality steel. In addition, argon blowing and stirring can accelerate the reduction of oxide in slag, which is beneficial to the recovery of valuable alloy elements such as chromium, molybdenum and tungsten. The bottom blowing argon agitation runs through the whole refining process, that is to say, the argon blowing is started at the time of tapping of the primary furnace to prevent the blockage of the air permeable brick until the end of the refining process. The liquid steel is stirred by blowing argon in the way of first strong and then weak. The composition and temperature of the molten steel are uniformly stirred by bottom blowing argon to promote the floatation of deoxidation products, and remove some gases from the steel and purify the molten steel.

吹氩搅拌的另一作用是可以加速钢液中的温度与成分均匀,能精确调整复杂的化学组成,而这对优质钢又是必不可少的要求。此外,吹氩搅拌可加速渣中氧化物的还原,有利于回收铬、钼、钨等有价值的合金元素。底吹氩气搅拌贯穿于整个精炼过程,即在初炼炉出钢时就开始吹氩,以防止透气砖堵塞,直至精炼过程结束停止吹氩。按照先强后弱的搅拌方式吹氩搅拌钢液。通过底吹氩搅拌均匀钢液成分和温度,促进脱氧产物的上浮,脱除钢中的部分气体,从而纯净钢液。

Bottom blown argon plays an important role in LF treatment process. Bottom blowing argon has great influence on slagging, composition regulation, desulfurization, deoxidization and inclusion floatation. The stirring strength is also related to the slag content and molten steel temperature. The practice shows that after LF refining furnace arrives at the ladle lifting station, it is connected with argon blowing pipe to break the shell by side blowing. After breaking the shell, the flow rate slightly decreases, which is subject to the fact that the molten steel is not exposed. After arriving at the heating station, the medium strength argon flow rate is adopted for mixing according to the ventilation condition of the permeable brick. After the ladle is fed with Ca-Si wire out of the heating station, soft mixing is carried out. The change range of argon flow is about $350 \sim 420 \text{N} \cdot \text{m}^3/\text{min}$.

底吹氩在 LF 炉处理过程中发挥着重要作用。底吹氩对化渣、调节成分、脱硫、脱氧及夹杂物的上浮均有较大影响，搅拌强度与钢包带渣量和钢水温度也有较大关系。实践表明，LF 精炼炉到吊包工位后，接吹氩管进行旁吹破壳，破壳后流量稍微变小，以钢水不裸露为准。到加热工位后，根据透气砖透气情况，采取中等强度的氩流量进行搅拌。钢包出加热工位喂 Ca-Si 丝后，进行软搅拌。氩气流量变化范围约为 $350 \sim 420 \text{N} \cdot \text{m}^3/\text{min}$。

Several refining functions of LF furnace influence, depend on and promote each other. The reduction atmosphere in the furnace and the stirring of the steel slag under the heating condition improve the refining capacity of the synthetic slag and create an ideal steelmaking environment so as to improve the quality of the molten steel. The LF method is simpler than ASEA-SKF method and VaD method because it uses argon blowing instead of electromagnetic stirring and heating at atmospheric pressure instead of vacuum heating. Its metallurgical effect is basically the same as the former two methods. Through LF refining, the following results are achieved:

LF 炉的几个精炼功能互相影响，互相依存，互相促进。炉内的还原气氛以及加热条件下的钢渣搅拌，提高了合成渣的精炼能力，同时创造出一个理想的炼钢环境，从而提高钢液质量。由于 LF 法采用吹氩搅拌而不采用电磁搅拌，此外在大气压下加热，而不采用真空加热，因此 LF 法比 ASEA-SKF 法和 VAD 法都简单，其冶金效果与前两种方法基本相同。通过 LF 精炼，主要能达到以下效果：

(1) The temperature of molten steel can meet the requirements of continuous casting process.

(1) 钢液温度满足连铸工艺要求。

(2) The treatment time meets the requirements of multi furnace continuous casting.

(2) 处理时间满足多炉连浇要求。

(3) The fine adjustment of ingredients can ensure that the products have qualified ingredients and realize the lowest cost control.

(3) 成分微调能保证产品具有合格的成分，从而实现最低成本控制。

(4) The purity of molten steel can meet the quality requirements of products, especially with strong desulfurization function.

(4) 钢液纯净度能满足产品质量要求，并且具有强脱硫功能。

Task 2.2　LF Refining Slag Technology
任务 2.2　LF 精炼造渣技术

High basicity slag refining is an important operation of LF furnace. Lime, fluorite, etc. are added to the ladle in different proportion (such as 5∶1 or 4∶1), the amount of which is 1%~

2% of the liquid steel, making high basicity synthetic refining slag for desulfurization; and then ferrosilicon powder, calcium silicon powder, and aluminum powder (or carbon powder) are mixed in a certain proportion and directly added to the steel surface or added to the liquid steel by blowing method, forming the slag with good fluidity, with the composition (by mass) of $w(CaO) = 60\% \pm 5\%$, $w(SiO_2) = 10\% \pm 5\%$, $w(Al_2O_3) = 30\% \pm 5\%$ or $w(CaO) = 60\%$, $w(Al_2O_3) = 30\%$, $w(CaF_2) = 10\%$. In addition to the slag with high basicity in reductive atmosphere, the main factors affecting the desulfurization reaction also include the influence of stirring intensity of argon blowing. Argon blowing stirring at the bottom of the ladle can accelerate the diffusion of sulfur in the molten steel to the slag and steel interface, which can increase the desulfurization rate by 1.5~2.0 times and reach more than 90%.

造高碱度渣精炼是 LF 炉的重要操作。将石灰、萤石等按不同比例（如 5∶1 或 4∶1）分批加入钢包内，加入量为钢液量的 1%~2%，造高碱度合成精炼渣脱硫；然后用硅铁粉、硅钙粉和铝粉（或炭粉），按一定比例混合直接加入钢液面或采取喷吹方法加入钢液中，形成流动性良好的炉渣，其成分（质量分数）为：$w(CaO) = 60\% \pm 5\%$，$w(SiO_2) = 10\% \pm 5\%$，$w(Al_2O_3) = 30\% \pm 5\%$ 的还原渣系，或为：$w(CaO) = 60\%$，$w(Al_2O_3) = 30\%$，$w(CaF_2) = 10\%$ 的还原渣系。影响脱硫反应的主要因素除了还原性气氛下高碱度的炉渣之外，还有吹氩搅拌强度的影响，包底吹氩搅拌，加速钢液中的硫向渣、钢界面扩散，可使脱硫率提高 1.5~2.0 倍，达到 90% 以上。

LF refining process makes full use of slag washing method, and realizes refining effect by using the better fluidity of ladle slag, foaming submerged arc effect, desulfurization and inclusion absorption ability. The functions of ladle slag mainly include:

LF 精炼工艺充分利用渣洗手段，利用精炼过程钢包渣较好的流动性、发泡埋弧作用、脱硫及吸收夹杂物的能力实现精炼效果。钢包渣的作用主要包括：

(1) Heat preservation and insulation of molten pool to maintain the temperature of molten steel.

(1) 熔池保温隔热，维持钢水温度。

(2) Remove inclusions from steel.

(2) 从钢中去除夹杂。

(3) Protect molten steel from oxidation and control its chemical composition.

(3) 保护钢水免受氧化，控制钢水化学成分。

(4) Removal of harmful inclusions and improvement of steel properties.

(4) 脱除有害夹杂硫，改善钢的性质。

(5) Stable arc and submerged arc heating to protect the refractory from arc ablation.

(5) 稳定电弧和埋弧加热，使耐材免受弧光烧蚀。

(6) Ensure the use of refractory materials and minimize the erosion of refractory materials.

(6) 保证耐材使用，最大限度减少耐材侵蚀。

In the smelting process, the above requirements must be taken into account for the refining slag, and the type of refining slag is related to the steel type produced, the composition of molten steel entering the station, raw materials, the actual needs of each steelmaking plant, etc. For example, the converter production process of a steel plant, high efficiency production, requires the ladle furnace to only play the role of temperature regulation and buffering, while the submerged arc heating method used in the refining and slagging process is particularly important at this time. When refining steel grades in another steel plant furnace, desulfurization is especially required. In addition to conventional functions, refining slag needs to have good desulfurization capacity. The target slag system index of 100t LF refining is shown in Table 2-1.

冶炼过程中要求精炼渣必须考虑上述要求，而精炼渣的类型则与生产的钢种，以及进站的钢水成分、原料、各炼钢厂的实际需要等因素有关。例如，某钢厂转炉生产流程，高效率生产，要求钢包炉只起调温、缓冲的作用，而此时精炼造渣过程运用的埋弧加热手段尤为重要。另一钢厂在进行炉精炼钢种时，特别要求脱硫，精炼渣除具备常规功能外，需要有良好的脱硫能力。其100t LF 精炼目标渣系指标见表2-1。

Table 2-1　100t of LF refining slag system
表2-1　某厂100t LF 精炼炉渣系

\multicolumn{8}{c}{Components（mass fraction）/%}							
CaO	SiO_2	Al_2O_3	MgO	FeO	MnO	CaF_2	$R=CaO/SiO_2$
44~45	10~20	15~20	5~10	<0.8	<0.3	7~12	2.9~3.6

2.2.1　Properties of Synthetic Slag

2.2.1　合成渣的性质

In order to achieve the purpose of refining molten steel, the synthetic slag must have high basicity, high reducibility, low melting point and good fluidity. In addition, it must have suitable density, diffusion coefficient, surface tension and conductivity.

为了达到精炼钢液的目的，合成渣必须具有较高的碱度、高还原性、低熔点和良好的流动性。此外，还要具有合适的密度、扩散系数、表面张力和导电性等。

2.2.1.1　Composition of Synthetic Slag

2.2.1.1　合成渣成分

In order to obtain the best refining effect, the synthetic slag is required to have the corresponding physical and chemical properties, and the composition of the slag is the decisive factor of the physical and chemical properties of the slag. The synthetic slag mainly includes $CaO-Al_2O_3$ system, $CaO-SiO_2-Al_2O_3$ system, $CaO-SiO_2-CaF_2$ system, etc. Among them, the mass fraction of each component is CaO: 45%~60%, MgO: 6%~10%, Al_2O_3: 12%~16%. SiO_2: 16%~20%.

为了取得最佳的精炼效果，要求合成渣具备相应的物理化学性质，而炉渣的成分是由炉渣的物理化学性质决定。合成渣主要包括：$CaO-Al_2O_3$ 系，$CaO-SiO_2-Al_2O_3$ 系，$CaO-SiO_2-CaF_2$ 系等。其中，合成渣中各物质含量（质量分数）为：CaO 45%~60%，MgO 6%~

10%, Al_2O_3 12%~16%, SiO_2 16%~20%。

2.2.1.2 Melting Point of Synthetic Slag

2.2.1.2 合成渣熔点

When the molten steel is refined with the synthetic slag in the ladle, the liquid slag is generally used. So the melting point of the synthetic slag should be lower than that of the molten steel to be refined. The melting point of molten steel can be calculated by

在钢包内用合成渣精炼钢液时，一般应用液态渣，因此合成渣的熔点应当低于被精炼钢液的熔点。钢液的熔点的计算公式为

$$T_f = 1538 - \Sigma \Delta T_j [j\%] \quad (2-3)$$

Where T_f——the approximate melting point of steel containing elements in ℃;

ΔT_j——the melting point of pure iron decreases when the element in steel increases by 1% in ℃;

[j%]——the mass percentage of element J in steel in %;

1538——the melting point of pure iron in ℃.

式中 T_f——含元素钢的近似熔点,℃;

ΔT_j——钢中元素增加1%时纯铁熔点的降低值,℃;

[j%]——钢中j元素含量（质量分数）,%;

1538——纯铁的熔点,℃。

2.2.1.3 Fluidity of Synthetic Slag

2.2.1.3 合成渣流动性

The fluidity of synthetic slag is one of the important factors that affect the refining effect and the emulsification degree of slag in molten steel. Under the same temperature and mixing condition, increasing the fluidity of synthetic slag can reduce the average diameter of emulsified slag drop. In $CaO-Al_2O_3$ slag system, with the increase of SiO_2 content, the viscosity decreases and the fluidity becomes better, thus increasing the contact area of slag steel. For most of the synthetic slag, its viscosity is less than 0.2 Pa·s at the steelmaking temperature.

合成渣要求有较好的流动性，其流动性影响精炼的效果。在相同温度和混冲条件下，提高合成渣的流动性可以减小乳化渣滴的平均直径。在$CaO-Al_2O_3$渣系中，随着SiO_2含量的提高，其黏度降低，流动性变好，从而增大了渣钢接触界面积。对于大部分合成渣，在炼钢温度下其黏度小于0.2Pa·s。

2.2.1.4 Surface Tension of Synthetic Slag

2.2.1.4 合成渣表面张力

Surface tension is also an important parameter affecting refining effect. In the refining process of synthetic slag, although the interfacial tension between steel-slag and slag-inclusions (which affects the ability of slag to absorb and assimilate non-metallic inclusions), the magnitude of the interfacial tension is directly related to the surface tension of each phase.

表面张力也是影响精炼效果的一个较为重要的参数。在合成渣精炼过程中，虽然直接起作用的是钢渣之间的界面张力和渣与夹杂物之间的界面张力（影响渣滴吸附和同化非金属夹杂的能力），但是界面张力的大小与每一相的表面张力直接有关。

2.2.1.5　Reducibility of Synthetic Slag

2.2.1.5　合成渣还原性

According to the task of refining, it is required that the nature of synthetic slag is reductive slag, and the content of FeO in the slag is very low, generally lower than 0.3% (by mass).

根据精炼要完成的任务，要求合成渣的性质是还原性渣，渣中 FeO 含量（质量分数）都很低，一般应低于 0.3%。

2.2.2　Removal of Inclusions

2.2.2　夹杂物的去除

In the refining process of synthetic slag, the molten steel is stirred by blowing argon at the bottom of the ladle at the same time. The argon blown in drives the molten steel to move up and down, so that the inclusions in the molten steel float up to the steel surface absorbed and assimilated by the molten slag (The inclusions are absorbed by the molten slag and dissolved in the slag drops) and enter the slag phase. Because the interfacial tension between slag and inclusions is much less than that between molten steel and inclusions, inclusions in steel are easily absorbed by slag. The slag used for refining synthetic slag is oxide melt, and most of the inclusions are also oxide, so the inclusions adsorbed by the slag are easy to dissolve in the slag.

在合成渣精炼过程中，同时进行钢包底吹氩搅拌钢液，吹入的氩气驱动钢液上下运动，使钢液内的夹杂物上浮到钢液面被熔渣吸附、同化（夹杂被熔渣吸附并溶解于渣滴中的过程称作同化）而进入渣相。由于渣和夹杂间的界面张力远小于钢液与夹杂间的界面张力，所以钢中夹杂很容易被熔渣所吸附。合成渣精炼所用的熔渣均是氧化物熔体，而夹杂物多数也是氧化物，所以被渣吸附的夹杂比较容易溶解于渣中。

On the one hand, Deoxidizer should be used to reduce the dissolved oxygen in molten steel and the content of slag (FeO+MnO) in the LF refining process; On the other hand, stirring measures should be taken to remove the deoxidized products. When the acid soluble aluminum in steel reaches 0.03%~0.05%, the deoxidation of molten steel is complete. At this time, the dissolved oxygen in steel almost changes into Al_2O_3. The essence of deoxidation of molten steel is the removal of oxide inclusions.

LF 精炼过程一方面要用脱氧剂最大限度地降低钢液中的溶解氧，同时进一步减少渣中（FeO+MnO）的含量；另一方面要采取搅拌措施使脱氧产物上浮去除。用强脱氧元素铝脱氧，钢中的酸溶铝达到（质量分数）0.03%~0.05%时，钢液脱氧完全，这时钢中的溶

解氧几乎都转变成 Al_2O_3，钢液脱氧的实质是钢中氧化物夹杂去除问题。

After aluminum deoxidization, the dissolved oxygen in molten steel increases with the refining process.

铝脱氧后，钢液中的溶解氧随着精炼过程的进行有所升高。

Acid soluble aluminum refers to the content of AlS in steel obtained by acid soluble method. In LF refining process, the oxidation of acid soluble aluminum in molten steel is mainly caused by the oxidation of SiO_2, MnO, FeO, Cr_2O_3 in slag and atmosphere.

酸溶铝是指用酸溶法得到的钢中的 AlS 含量。在 LF 精炼过程中，引起钢液中酸溶铝氧化的主要有渣中 SiO_2、MnO、FeO、Cr_2O_3 以及大气的氧化。

In LF refining, the refining time should be ensured to make oxide inclusions fully float.

LF 精炼要保证精炼时间，使氧化物夹杂充分上浮。

2.2.3 Desulfurization Process

2.2.3 脱硫工艺

Desulfurization is the main purpose of refining synthetic slag, which can reduce [S] to less than 0.005% and [O] to less than 0.002%. If operated properly, it can generally remove [S] by about 50%~80%. In order to maximize the desulfurization, the basicity of the synthetic slag must be high, so as to enhance the sulfuration ability, improve the fluidity of the slag and avoid damaging the refractories. In order to desulfurate, the molten steel must be deoxidized with aluminum before tapping. In order to achieve the full desulfuration effect, the residual aluminum content should be more than 0.02%. The amount of slag in tapping process should be reduced as far as possible, because the (FeO) in the slag can significantly reduce the desulfurization rate, reduce the amount of slag in the furnace to the minimum and carry out slag modification treatment.

脱硫是合成渣精炼的主要目的，可以使 [S] 降到 0.005%（质量分数）以下，[O] 降到 0.002%（质量分数）以下；如果操作得当，一般可以去除 [S] 约 50%~80%（质量分数）。为了最大限度地脱硫，合成渣的碱度必须高，以使硫化能力增强，渣的流动性良好，不损坏耐火材料。为了脱硫，出钢前钢液必须先用铝脱氧；为了达到充分脱硫效果，需要残余铝量在 0.02%（质量分数）以上。应尽量减少出钢过程的下渣量，因渣中的（FeO）能明显降低脱硫率，使炉内下渣量减到最低限度并进行渣改性处理。

In the LF furnace, high basicity and strong reductive synthetic slag can be produced. With the stirring of ladle bottom blowing argon, the extremely superior thermodynamic and dynamic conditions of desulfurization can be created. It can effectively desulfurize and is suitable for the production of low sulfur steel and ultra-low sulfur steel. The technological conditions for smelting ultra-low sulfur steel in LF furnace are as follows:

在 LF 炉内可以造高碱度强还原性合成渣，配合钢包底吹氩搅拌，创造极为优越的脱硫热力学和动力学条件，能够有效地脱硫，因此适合于低硫钢、超低硫钢的生产。使用 LF 炉冶炼超低硫钢的工艺条件为：

(1) They slag composition is controlled and the slag basicity is improved. Therefore, the

content of SiO_2 in slag should be controlled below 10%, and the best level is 5%. For special occasions, more basicity components such as BaO, Na_2O and Li_2O can be added.

(1) 控制炉渣成分，提高炉渣碱度。此外，炉渣中 SiO_2 的含量（质量分数）要控制在10%以下，最好达到5%的水平。为了适应特殊场合，可以添加 BaO、Na_2O、Li_2O 等碱度更高的组元。

(2) The deoxidation of slag and molten steel is strengthened. By adding diffusion deoxidizer to slag, the content of (FeO +MnO) in slag can reach 1% or even less than 0.5%. The content of acid soluble aluminum in steel is controlled so that the oxygen activity in molten steel is controlled below 1.0×10^{-3}.

(2) 强化对炉渣和钢水的脱氧。向炉渣中加入扩散脱氧剂，使渣中（FeO+MnO）含量（质量分数）达到1%，甚至0.5%以下。控制钢中酸溶铝含量，使钢水中氧活度控制在 1.0×10^{-3} 以下。

(3) High refining temperature and good bottom blowing argon stirring process are also important desulfurization process conditions.

(3) 较高的精炼温度和良好的底吹氩搅拌工艺也是脱硫工艺的重要条件。

(4) The original sulfur content of slag and molten steel will be limited, and the corresponding slag quantity will be ensured at the same time. If necessary, the slag replacement operation can be carried out.

(4) 对炉渣和钢水的原始硫含量进行限制，同时保证相应的渣量，必要时可进行换渣操作。

LF furnace uses the white slag (reducing slag) produced in the refining process to refine, realize desulfurization, deoxidization, and produce ultra-low sulfur and low oxygen steel. Therefore, the white slag refining is the core of LF furnace process operation. Refining process requirements are as follows:

LF 炉利用精炼过程产生的白渣（还原性渣）进行精炼，实现脱硫、脱氧、生产超低硫和低氧钢，因此白渣精炼是 LF 炉工艺操作的核心。精炼过程要求如下：

(1) Slag blocking by tapping, controlled slag amount is no more than 5kg/t.

(1) 出钢挡渣，控制下渣量不小于 5kg/t。

(2) Ladle slag will be modified to control $R \geq 2.5$ and $w(TFe+MnO) \leq 3.0\%$ in slag.

(2) 钢包渣改质，控制 $R \geq 2.5$，渣中 $w(TFe+MnO) \leq 3.0\%$。

(3) Generally, Al_2O_3-CaO-SiO_2 system slag is used for white slag refining, with $R \geq 4$ controlled and $w(TFe+MnO) \leq 1.0\%$ in slag.

(3) 白渣精炼，一般采用 Al_2O_3-CaO-SiO_2 系炉渣，控制 $R \geq 4$，渣中 $w(TFe+MnO) \leq 1.0\%$。

(4) Control the atmosphere in the furnace to be weak oxidization to avoid slag reoxidation.

(4) 控制炉内气氛为弱氧化性，避免炉渣再氧化。

(5) Proper agitation can avoid the exposed steel surface and ensure the high mass transfer speed in the molten pool.

(5) 适当搅拌，避免钢液面裸露，并保证熔池内具有较高的传质速度。

2.2.3.1 Desulfurization Mechanism of Refining Slag

2.2.3.1 精炼渣脱硫机理

From the thermodynamic point of view, high temperature is conducive to desulfurization. And higher temperature can result in better kinetic conditions and accelerate desulfurization reaction. In addition, the effect of FeO and MnO in ladle slag on the desulfurization capacity of the slag can not be ignored, and should be controlled in a lower concentration range as far as possible. It is concluded that the removal of sulfur in steel is related to the following factors:

从热力学的角度讲,温度高有利于脱硫反应。而且较高的温度可以造成更好的动力学条件,从而加快脱硫反应。另外,钢包渣中的 FeO、MnO 对脱硫渣的脱硫能力有影响,应该尽量控制在较低的浓度范围。钢中硫的去除与以下因素有关:

(1) Mixing degree of steel and slag (钢同渣的混合程度).
(2) Slag quantity (渣量).
(3) Temperature (温度).
(4) Oxygen content in steel (钢中氧含量).
(5) Oxidation degree of slag (炉渣氧化程度).
(6) Viscosity of slag (炉渣的黏度)

The desulfurization reaction is as follows:

其脱硫反应为:

$$[S]+(CaO)=\!\!=\!\!=(CaS)+[O] \tag{2-4}$$

2.2.3.2 Effect of Slag Components on Desulfurization

2.2.3.2 炉渣各组分对脱硫的影响

In order to desulfurate the molten steel, it is necessary to fully deoxidize the molten steel. At this time, the aluminum content in steel should be higher than 0.02% (by mass). At this time, it can be ensured that the dissolved oxygen is not higher than 2~4mg/L. It is necessary to ensure the high basicity and strong reducibility of the slag (The free CaO content in the slag is high), and the $w(FeO+MnO)$ in the slag is sufficiently low, generally less than 0.5%. Too much SiO_2 will reduce the desulfurization capacity of slag, but it can reduce the melting point of slag, make slag participate in the reaction as soon as possible, and play a favorable role in desulfurization. As long as it does not exceed 5%, it will not cause adverse impact on desulfurization.

要使钢水脱硫,首先必须使钢水充分脱氧。此时钢中的铝含量(质量分数)应当高于 0.02%。这时可以保证溶解氧不高于 2~4mg/L。脱硫应保证炉渣的高碱度、强还原性(渣中自由 CaO 含量要高);渣中 $w(FeO+MnO)$ 要充分低,一般小于 0.5%。过多的 SiO_2 会降低炉渣的脱硫能力,但是却可以降低炉渣的熔点,使炉渣尽快参加反应,起到对脱硫有利的作用,即只要不超过 5%就不会对脱硫造成不利影响。

Effect of Basicity on Desulfuration Performance of Refining Slag
碱度对精炼渣脱硫性能的影响

The basicity of refining slag has great influence on the desulfurization and deoxidization of refining process. The increase of basicity can reduce the equilibrium oxygen and increase the distribution ratio of sulfur between slag and steel. The results showed that when $R<3.0$, alkalinity increased and L_S increased, while when $R>3.0$, R increased and L_S decreased. In each desulfurization slag system, CaO plays an important role in desulfurization, and its content will directly affect the desulfurization effect. With the increase of CaO, [S] will decrease, but when $w(CaO)>60\%$, the increase of CaO content will reduce the desulfurization effect. As the CaO content is too high, there will be solid particle precipitation in the slag, which makes the slag appear heterogeneous phase. The slag viscosity increases and the fluidity becomes poor, thus affecting the dynamic conditions of desulfurization.

精炼渣碱度对精炼过程的脱硫、脱氧均有较大影响。碱度提高可使钢中平衡氧降低,同时提高硫在渣钢间的分配比。实验表明,当 $R<3.0$ 时,碱度增加,L_S 随之增加;当 $R>3.0$ 时,继续增加 R,L_S 下降。在各脱硫渣系中,对脱硫起主要作用的组元是 CaO,其含量的高低将直接影响脱硫效果的好坏。随着 CaO 的提高,[S] 降低,但当 $w(CaO)>60\%$ 以后,CaO 含量的提高使脱硫效果降低。由于 CaO 含量过高后,渣中会有固相质点析出,使熔渣出现非均相,炉渣黏度上升,流动性变差,从而影响了脱硫的动力学条件。

The alkalinity of foreign LF refining slag is higher. Sometimes the content of CaO in slag is as high as 65% (by mass); and the alkalinity of domestic LF refining slag is mostly at the middle (2.2~3.0) and low (1.6~2.2). But at present, many steel plants adopt high alkalinity process. There are different requirements for the final sulfur content of the products, and the alkalinity of different plants and different steel grades is quite different in actual production. For example, as shown in Tables 2-2 and 2-3 for slag composition before and after LF slagging in a domestic factory, the highest desulfurization rate of the slag after LF refining can reach 66.67%, with an average of 62.5%.

国外 LF 精炼渣碱度较高,有时渣中 CaO 含量高达 65%(质量分数);国内 LF 精炼渣碱度多数处于中(2.2~3.0)、低(1.6~2.2)水平。但目前许多钢厂均采用高碱度工艺技术路线。产品最终硫含量要求不同,实际生产中各厂及不同钢种碱度差别较大。例如,国内某厂生产 LF 造渣前后炉渣成分见表 2-2 和表 2-3,利用 LF 精炼该渣洗脱硫率最高可达 66.67%,平均为 62.5%。

Table 2-2　Slag composition before LF slag making

表 2-2　LF 造渣前炉渣成分

炉次 No.	Components (mass fraction)/% 化学成分(质量分数)/%									
	S	FeO	MnO	P_2O_5	CaO	MgO	SiO_2	Al_2O_3	R	FeO+MnO
5D8530	0.029	1.7	11.5	0.017	31.08	10.43	9.66	35.1	3.22	13.2
4D7889	0.032	1.41	8.28	0.018	33.62	7.77	9.23	39.24	3.64	9.69

Continued Table 2-2

炉次 No.	Components (mass fraction)/% 化学成分（质量分数）/%									
	S	FeO	MnO	P_2O_5	CaO	MgO	SiO_2	Al_2O_3	R	FeO+MnO
5D8531	0.032	2.14	9.11	0.025	34.75	9.74	10.17	32.19	3.42	11.25
4D7890	0.025	2.02	15	0.016	25.41	16.7	10.11	30.03	2.51	17.02
6D8420	0.034	3.23	4.03	0.018	30.41	11.47	9.66	39.08	3.15	7.26
6D8421	0.03	2.14	10.87	0.018	29.13	16.55	9.03	35.63	3.23	13.01
Average 平均	0.03	2.11	9.80	0.02	30.73	12.11	9.64	35.21	3.20	11.91

Table 2-3 Slag composition after LF slagging
表 2-3 LF 造渣后炉渣成分

炉次 No.	Components (mass fraction)/% 化学成分（质量分数）/%									
	S	FeO	MnO	P_2O_5	CaO	MgO	SiO_2	Al_2O_3	R	FeO+MnO
5D8530	0.15	1.14	0.78	0.016	46.92	8.35	6.23	35.39	7.53	1.92
4D7889	0.1	0.62	0.45	0.016	45.98	5.88	7.81	38	5.89	1.07
5D8531	0.065	1.65	0.35	0.019	41.74	9.35	7.34	39.26	5.69	2
4D7890	0.13	0.74	0.76	0.021	41.35	9.07	6.18	38.84	6.69	1.5
6D8420	0.15	1.46	0.67	0.021	42.42	8.49	6.82	40.18	6.22	2.13
6D8421	0.1	1.45	1.93	0.015	47.5	8.85	8.68	33.05	5.47	3.38
Average 平均	0.12	1.18	0.82	0.02	44.32	8.33	7.18	37.45	6.25	2.00

Effect of CaF_2 on Desulfurization
CaF_2 对脱硫的作用

CaF_2 does not have the function of desulfuration. Its main function is to reduce the melting point of desulfuration slag and improve the fluidity of desulfuration slag. CaF_2 and CaO form a series of eutectic substances with low melting point to help melt slag. With the progress of desulfuration reaction, there will be the formation of CaS solid phase in the slag gold interface, and the presence of CaS solid phase will prevent the continuous process of desulfuration reaction and reduce the amount of liquid phase.

CaF_2 本身不具备脱硫作用，其主要作用是降低脱硫渣的熔点，改善脱硫渣的流动性。CaF_2 与 CaO 形成一系列低熔点共晶物来助熔化渣。随着脱硫反应的进行，渣—金界面将有 CaS 固相形成，而 CaS 固相的存在，阻止了脱硫反应的继续进行，从而使液相量减少。

Adding CaF_2 into slag is beneficial to the destruction of solid phase, the increase of liquid phase and the improvement of desulfurization conditions. However, when the content of CaF_2 in the slag reaches enough to prevent the formation of CaS solid, increasing CaF_2 will cause the CaO in the slag to be diluted and the effective CaO concentration will be reduced, which is not conducive to desulfurization.

渣中加入 CaF_2 有利于固相的破坏,使液相量增加,从而改善脱硫条件。但当渣中 CaF_2 含量达到足以阻止 CaS 固体形成时,若继续增加 CaF_2,则会造成渣中 CaO 被稀释,使有效 CaO 浓度降低,不利于脱硫。

There is no consensus on the optimal amount of CaF_2. The content of CaF_2 in refining slag fluctuates from 7.5% to 30% (by mass). In a word, alkalinity should be taken as the main reference factor in the design of synthetic refining slag system, and appropriate addition should be made on the premise of considering the fluidity of slag system, consumption of refractory and the requirements of environmental protection.

对于 CaF_2 的最佳加入量,目前并没有统一的观点。综合前人的研究成果,合成精炼渣中 CaF_2 含量(质量分数)波动在 7.5%~30% 之间。总之,进行合成精炼渣系设计时应以碱度作为主要参考因素,并在考虑渣系流动性、耐材消耗以及符合环保要求的前提下进行适量添加。

Effect of MgO in Slag on Desulfurization
渣中 MgO 对脱硫的影响

MgO is an alkaline material, which has a certain binding capacity with sulfur, but it is not as good as CaO. Its main function is to reduce the activity of SiO_2 in slag and to improve the activity of CaO, so as to improve the distribution coefficient L_s of sulfur between slag steels. But the disadvantageous factor for desulfurization is that MgO can improve the melting point of slag. Especially when the MgO content is more than 6.0%~8.0% (by mass), the slag will thicken rapidly, which is not conducive to the reaction. Generally, for the CaO-SiO_2-MgO-Al_2O_3 quaternary slag system, when the Al_2O_3 content is 15%~25% (by mass), the slag with $w(MgO)>10\%$ (by mass) enters the two-phase region. The appropriate MgO content in refining slag can protect the magnesia lining and reduce the melting process from lining to slag.

MgO 为碱性物质,与硫具有一定的结合能力,但不如 CaO。其主要作用是降低渣中 SiO_2 的活度,提高 CaO 的活度,从而提高硫在渣钢之间的分配系数 L_s。但 MgO 提高渣的熔点对脱硫不利,特别是当 MgO 含量(质量分数)大于 6.0%~8.0% 时,渣迅速稠化,不利于反应进行。一般对于 CaO-SiO_2-MgO-Al_2O_3 四元渣系,当 Al_2O_3 含量(质量分数)为 15%~25% 时,$w(MgO)>10\%$ 的熔点即进入两相区。精炼渣中适量的 MgO 含量可以起到保护镁质炉衬的作用,减少炉衬向熔渣的熔解过程。

The Role of Al_2O_3 in Refining Slag
Al_2O_3 在精炼渣中的作用

According to the ternary phase diagram of CaO-SiO_2-Al_2O_3, the melting temperature of slag decreases with the increase of Al_2O_3 content in a certain range. Increasing Al_2O_3 content in slag can promote slag formation and rapid slag formation. The rich Al_2O_3 reduction slag has the characteristics of loose and foam, and the reaction area is large, which can increase the activity of oxygen ions in the slag. Because of the amphoteric characteristics of Al_2O_3, the basicity of slag can be properly increased and the activity coefficient of sulfur ion in slag can be reduced. Meanwhile, the increase of its content has little effect on the viscosity of slag, which can ensure that slag has high basicity and good fluidity. It can also improve the heating speed of molten pool and the mass transfer coefficient of sulfur in molten pool, which is conducive to the increase of the distribution ratio of sulfur between slag and steel, and the increase of the mass transfer coefficient of sulfur in molten pool desulfurization speed. On the other hand, in the reduction slag rich in Al_2O_3, the melting point of calcium aluminate is low, the adhesion to the molten steel is large, and it can float out of the molten steel quickly. The calcium aluminate can also be well wetted by CAS, and float out of the molten steel together, so as to reduce the inclusion content in the steel.

根据 $CaO-SiO_2-Al_2O_3$ 三元相图，随着渣中 Al_2O_3 含量在一定范围内的提高，渣的熔化温度降低。提高渣中的 Al_2O_3 含量，能够促进化渣，进行快速造渣。富含的 Al_2O_3 还原渣有疏松和泡沫特性，反应面积大，可提高渣中的氧离子活度。由于 Al_2O_3 两性特征，故能适当提高熔渣碱度，降低熔渣中硫离子的活度系数。同时，其含量的提高对炉渣黏度影响较小，能保证炉渣有高的碱度和良好的流动性；还能提高熔池的升温速度，提高硫在熔池中的传质系数，这些都有利于提高硫在渣钢间的分配比，提高脱硫速度。另一方面，Al_2O_3 富含的还原渣中，铝酸钙熔点低，对钢液的黏附力大，能迅速从钢液中浮出；且 CaS 能很好地润湿铝酸钙，并与铝酸钙一起浮出钢液，从而降低钢中夹杂物含量。

Effect of FeO+MnO

FeO+MnO 的影响

The oxidizability of slag is the main factor affecting the desulfurization effect of slag, which is affected by the content of FeO+MnO. When the content of FeO+MnO is more than 1% (by mass), the desulfurization efficiency will decrease obviously. When the content of FeO+MnO in slag is less than 1% (by mass), better metallurgical effect can be obtained. When the FeO content is less than 1% (by mass), the desulfurization rate can reach 86%, and the sulfur distribution ratio can reach 120~150.

炉渣氧化性是影响炉渣脱硫效果的主要因素，是通过 FeO+MnO 的含量来影响脱硫效果。当 FeO+MnO 含量（质量分数）>1%时，脱硫效率将明显下降；当渣中 FeO+MnO 含量（质量分数）<1%，冶金效果较好，脱硫率可达 86%，硫的分配比为 120~150。

Effect of Slag Physical Properties on LF Desulfurization

炉渣物理性质对 LF 脱硫的影响

The physical properties of slag mainly include density, viscosity, melting temperature, surface tension and conductivity. Compared with desulfurization, viscosity and melting temperature are the most important factors. The effect of slag physical properties on LF desulfurization are as follows：

炉渣的物理性质包括的内容较多，主要有密度、黏度、熔化温度、表面张力和电导率等。相对于脱硫而言，影响最大的因素是黏度和熔化温度。以下分别介绍炉渣物理性质对 LF 脱硫的影响：

(1) The influence of slag viscosity: The viscosity of slag is the main factor affecting the desulfurization reaction of Slag—Steel interface. The mass transfer rate in liquid phase is inversely

proportional to the viscosity of slag. If the viscosity of slag is too large, the dynamic condition of desulfurization will be worsened, which will make desulfurization difficult. Increasing the fluidity of slag can reduce the average diameter of emulsified slag drop, the contact area of slag steel is increas, and the desulfurization is promoted. If the viscosity is too small, the ability of slag to penetrate refractory is strong, which will increase the loss of refractory. At the same time, it is not conducive to the realization of submerged arc operation in the refining furnace.

（1）炉渣黏度的影响：炉渣黏度是影响渣—钢界面脱硫反应的主要因素，液相中的传质速率与熔渣的黏度呈反比。若炉渣黏度过大，则恶化了脱硫的动力学条件，造成脱硫困难。提高炉渣的流动性，可以减小乳化渣滴的平均直径，增大渣钢接触面积，从而促进脱硫。若黏度过小，炉渣对耐火材料的渗透能力强，会造成耐火材料损耗增加，同时在精炼炉中也不利于实现埋弧操作。

（2）The influence of slag melting temperature: Whether the desulfuration process, which can be carried out normally, is often related to the melting temperature of slag. At a certain furnace temperature, the lower the melting temperature of slag, the higher the superheat, the better the fluidity and the faster the desulfurization reaction between slag and steel. The main factors affecting the melting performance are temperature, basicity, and the contents of SiO_2, Al_2O_3 and MgO.

（2）炉渣熔化温度的影响：炉外精炼脱硫过程能否正常进行，往往与炉渣的熔化温度有关。在一定炉温下，炉渣的熔化温度越低，过热度越高，流动性越好，渣—钢间脱硫反应就越快。熔化性能的影响因素主要有温度、碱度、SiO_2 含量、Al_2O_3 含量和 MgO 含量。

Effect of Smelting Process Conditions on Desulfurization
冶炼工艺条件对脱硫的影响

The effects of smelting process conditions on desulfurization are as follows:
冶炼工艺条件对脱硫的影响包括：

（1）The influence of LF operating temperature: According to thermodynamics, desulfurization is endothermic reaction, so high temperature is favorable for desulfurization. The effect of temperature on desulfurization is mainly in dynamics. With the increase of temperature, the viscosity of slag decreases, and the fluidity of steel slag is improved, so the dynamics condition of desulfurization is improved. With the increase of temperature, the diffusion speed of Slag—Steel will be accelerated, which is also conducive to desulfurization.

（1）LF 操作温度的影响：从热力学方面分析，脱硫反应属于吸热反应，因而高温有利于脱硫。温度对脱硫的影响主要在动力学方面，随着温度的升高，炉渣黏度下降，改善钢渣流动性，从而提高脱硫的动力学条件。温度升高，使渣—钢间的扩散速度加快，有利于脱硫。

（2）The influence of slag quantity: When the composition of slag is fixed, the desulfurization efficiency mainly depends on the amount of slag. When considering the amount of synthetic slag, it is necessary to pay attention to the influence of the amount of oxide slag, deoxidation products and erosion of ladle lining. The process of refining should not only meet the requirements of reasonable slag composition, but also meet the requirements of slag quantity. A certain temperature drop shall be allowed at the same time. When smelting Ultra-low sulfur steel, the amount of slag recommended in the literature is $12\sim20kg/t_{steel}$.

（2）渣量的影响：当熔渣组成一定时，其脱硫效率主要取决于渣量。在考虑合成渣用量时，必须注意进入钢包内的氧化渣量、脱氧产物及包衬侵蚀量等的影响。精炼过程既要满足合理的熔渣组成要求，又要满足对渣量的要求，同时要允许一定温降。冶炼超低硫钢时，文献推荐的渣量为 $12\sim20kg/t_{钢}$。

（3）The influence of total oxygen content (dissolved oxygen) in steel: The oxygen level of molten steel is one of the important factors affecting the sulfur equilibrium distribution ratio, and the low oxygen level of molten steel is very necessary for the smelting of extremely low sulfur steel. Sulfur and oxygen belonging to the same group of elements have similar properties, but the chemical properties of oxygen are more active than sulfur. The desulfurization reaction of slag shows that with the increase of oxygen content in steel, the desulfurization reaction will be inhibited, and measures should be taken to reduce the oxygen content in steel as much as possible. The oxygen content in slag also affects the dissolved oxygen content in steel. There is an equilibrium distribution of oxygen between slag and steel. When the oxidation of slag is high, the slag will supply oxygen to the steel and increase the dissolved oxygen in the molten steel.

（3）钢中全氧量（溶解氧）的影响：钢水氧位是影响硫平衡分配比的重要因素之一，低氧位的钢水对于极低硫钢的冶炼是十分必要的。硫和氧属同族元素，具有相似的性质，但氧的化学性质比硫更为活泼，因此要得到极低硫钢必须先得到超低氧钢。炉渣脱硫反应表明，提高钢中氧含量将抑制脱硫反应的进行，因此应尽量采取措施降低钢中氧含量。渣中氧含量也会影响钢中的溶解氧含量。在渣/钢间存在氧的平衡分配，当炉渣氧化性较高时，炉渣会向钢中供氧，增加钢液中的溶解氧量。

（4）The influence of bottom blowing argon stirring: The stirring of bottom blowing argon is the main factor affecting the desulfurization speed. The main function of stirring is that it can improve the mass transfer coefficient, enlarge the contact area of slag/steel, and promote the floating of inclusions. If the amount of argon blowing is too large, the fluctuation of slag layer will be too large, even the liquid steel will be exposed, resulting in secondary oxidation, which will improve the oxygen level of molten steel and reduce the sulfur balance distribution of slag steel. Some documents think that the desulfurization rate increases with the increase of argon blowing rate. The greater the argon supply, the stronger the stirring of molten steel, the better the mixing degree of steel slag and the desulfurization reaction kinetic conditions, and the larger the desulfurization reaction area.

（4）底吹氩搅拌的影响：底吹氩气搅拌是影响脱硫速度的主要因素。搅拌的作用主要在于提高传质系数，扩大渣/钢接触面积，同时促进夹杂物上浮。如果吹氩量太大，会造成渣层波动过大，甚至钢液裸露，造成二次氧化，提高钢水氧位，降低渣钢硫平衡分配。部分文献认为，吹氩量增大，脱硫率提高。供氩量越大，钢水搅拌越强烈，钢渣混合程度越好，脱硫反应界面积越大，则越有利于脱硫反应动力学条件的改善。

In the actual production, the amount of argon blowing should be increased properly within the range of ensuring that the steel surface is exposed and the allowable temperature drop is not caused by blowing the slag layer. For example:

在实际生产中,应在保证不吹开渣层造成钢液面裸露和允许温降的范围内,适当加大吹氩量。例如:

1) The argon blowing condition of pipeline steel X70 LF produced by plant a in China. There are two mixing methods for 100t ladle:

1) 国内 A 厂生产管线钢 X70 LF 吹氩工况。100t 钢包分两种搅拌方式:

① Argon blowing capacity of strong stirring (强搅拌吹氩量) is 400~550NL/min;

② Argon blowing capacity of weak stirring (弱搅拌吹氩量) is 250~300NL/min.

Note: The strong agitation will be adopted with high temperature.

注:高温脱硫时采用强搅拌。

(2) The X65 LF refining argon blowing process of plant B in China:

(2) 国内 B 厂 X65 LF 精炼吹氩工艺:

1) LF incoming argon blowing volume (LF 进站吹氩量) is 100NL/min;

2) During the heating period, argon blowing volume (升温期间吹氩量) is 200NL/min;

3) During desulfurization, argon blowing volume (脱硫期间吹氩量) is 400NL/min;

4) At the end of LF refining, the amount of soft argon blowing (LF 精炼结束软吹氩量) is 40NL/min;

5) The processing time (处理时间) is 40~50min;

6) The end temperature of LF refining (LF 精炼结束温度) is 1650℃.

With the use of high basicity synthetic slag for refining, the slag has high desulfurization capacity, and [S] can be reduced to less than 0.002%. By argon blowing, stirring, synthetic slag, and reducing atmosphere, the slag has good deoxidation and inclusion removal capacity, and almost all inclusions larger than 10m are removed. The dissolved oxygen in steel can be reduced to less than 0.002%. Arc heating can accurately control the temperature of molten steel, adjust the alloy composition, and control the composition in a very narrow range. For example, the temperature deviation of molten steel can be controlled within 5℃ due to the stirring of argon blowing. In conclusion, the advantages of LF furnace in refining desulfurization are summarized as follows:

由于使用高碱度合成渣精炼,所以炉渣有较高的脱硫能力,[S] 可降到 0.002% 以下。通过吹氩搅拌,合成渣,还原性气氛,炉渣有很好的脱氧和去夹杂能力,大于 10m 的夹杂几乎全部去除。钢中溶解氧可降至 0.002% 以下,电弧加热可以精确控制钢液温度,调整合金成分,将成分控制在很窄的范围内。例如,由于吹氩搅拌可使钢液温度偏差控制在 5℃ 范围内,所以能把耐热钢、氮化钢等钢中 Al 和 Ti 控制在 0.05%(质量分数)左右。LF 炉在精炼过程脱硫的优势主要包括:

(1) It has good reducing atmosphere.

(1) 具有良好的还原性气氛。

(2) It can produce high basicity slag containing Al_2O_3.

(2) 能造含 Al_2O_3 的高碱度渣。

(3) The bottom of the ladle is agitated by argon blowing, the slag steel is sufficient, and the dynamic condition of desulfurization is superior.

(3) 包底吹氩搅拌,渣钢充分,脱硫的动力学条件优越。

(4) High temperature can meet the temperature required for desulfurization.

(4) 温度高,能够满足脱硫所需温度。

Task 2.3　LF Foam Slag Refining Process
任务 2.3　LF 泡沫渣精炼工艺

The research of slag foaming in metallurgical process is closely related to the development of iron and steel technology. In the open hearth steelmaking period, the phenomenon of foaming of metallurgical slag has attracted people's attention. The application of oxygen top blown converter and top bottom combined blown converter makes a great deal of research on the emulsification phenomenon and foam phenomenon of slag metal droplet bubble system and slag bubble system. The application of foam slag technology in electric furnace metallurgy is widely applied. After making use of slag foaming, it effectively shields the arc's strong radiation to refractory materials in the melting period. In the middle of 80s, the development and utilization of smelting reduction ironmaking technology made people have a further understanding of the phenomenon of foamed slag. According to the difference of metallurgical function, the foaming properties of slag in different reactors are quite different. Some metallurgical processes need to promote foaming of slag, while some metallurgical processes are opposite.

冶金过程中炉渣泡沫化的研究与钢铁工艺的发展密切相关。在平炉炼钢时期,冶金熔渣的泡沫化现象已经引起人们的注意。氧气顶吹转炉和顶底复吹转炉的应用使得冶金工作者对于炉渣—金属液滴—气泡体系、炉渣—气泡体系的乳化现象以及泡沫化现象进行了大量的研究,电炉炼钢中与长弧操作相适应的泡沫渣工艺也得到了广泛应用。电炉冶金中的泡沫渣技术的应用十分广泛,其利用了炉渣泡沫化后,能有效屏蔽电弧在熔清期对炉壁耐火材料的强烈辐射。20 世纪 80 年代中期熔融还原炼铁工艺的开发利用又使人们对泡沫渣的现象有了进一步认识。根据作用的差别,在不同的反应器中要求渣的泡沫化性质具有很大差别,有的冶金过程需要促进熔渣的泡沫化,而有的冶金过程则相反。

In the ladle refining furnace, using foam slag refining process can achieve submerged arc heating, stabilize the arc, improve the thermal efficiency of the refining process and the service life of the ladle for processing, protect the lining, and reduce the high temperature radiation of the heating arc on the ladle wall and the chance of the two oxidation of molten steel. Draw lessons from the technology of foam slag in EAF steelmaking to reduce the running cost of ladle refining furnace. In LF refining, the furnace wall of ladle is exposed to the radiation of electric arc. The working environment of refractory materials on the furnace wall is very bad, which causes the rapid consumption of ladle. The consumption of refractory materials accounts for more than half of the operation cost of ladle refining furnace. Based on its own characteristics, LF can not copy the foam slag process of electric furnace steelmaking. Because the low oxidation property of refining slag is the essential property of LF process, there is no large quantity of gas sources in the electric furnace. Therefore, it is necessary to use the technology which is not exactly the same as the electric furnace steelmaking to achieve the purpose of slag foaming.

在钢包精炼炉中使用泡沫渣精炼工艺,可以做到埋弧加热,稳定电弧,提高精炼过程的热效率;减少加热电弧对钢包炉壁的高温辐射,保护炉衬,提高处理用钢包的使用寿命;减少钢水的二次氧化机会。借鉴电炉炼钢的泡沫渣来降低钢包精炼炉运行成本的技术越来越被人们重视。LF精炼中钢包的炉壁暴露在电弧的辐射之中,炉壁耐火材料的工作环境极为恶劣,从而造成钢包快速消耗,耐火材料的消耗约占钢包精炼炉运行成本的一半以上。基于其自身的特点,LF不能照搬电炉炼钢的泡沫渣工艺。因为LF处理过程中精炼渣的低氧化性是其必须具备的性质,因而不存在电炉中存在大量气源,所以必须采用与电炉炼钢不完全相同的工艺达到熔渣泡沫化。

2.3.1 Foam Slag Metallurgical Effect

2.3.1 泡沫渣的冶金作用

Foaming slag is beneficial to the reaction of metallurgical, the improvement of steel quality and thermal efficiency, and the reduction of thermal erosion of lining. Depending on the resistance conversion of slag, the arc power is reduced under the same input power. The reduction of arc power and shielding of arc by slag are beneficial to reduce the heat load of furnace lining and the damage to slag line of furnace lining, and allow to change the power system. The short circuit arc extinguishing is reduced and avoided during melting. After slag foaming, and the conditions of arc region ionization are improved, and the conductivity of gas is increased with the arc resistance reduced. Under the same arc voltage, the arc length is increased, which ensures the stability of the arc, reduces or prevents short-circuit arc extinction, shortens smelting time and reduces electrode consumption. In addition, the foaming slag makes the slag steel interface expand, which is beneficial to the desulfurization reaction. As the arc has foam shielding, the partial pressure of nitrogen in the arc area is significantly reduced, which is conducive to the reduction of nitrogen uptake.

泡沫渣有利于冶金反应,提高钢质量,提高热效率,减少对炉衬的热侵蚀。依靠炉渣的电阻转换,在同样的输入功率下,减小了电弧功率。电弧功率减少和电弧被炉渣屏蔽,都有利于减少炉衬的热负荷,减少对炉衬渣线的损害,减少或避免熔化过程中的短路灭弧。渣子发泡后,弧区电离化条件得到改善,气体电导率增加,电弧电阻减少。在同样的电弧电压下,电弧长度增加,保证了电弧的稳定,减少或防止短路灭弧,缩短冶炼时间,降低电极耗损。另外,发泡的炉渣使渣钢界面扩大,有利于脱硫反应的进行。由于电弧有泡沫渣屏蔽,电弧区氮的分压显著降低,因此有利于减少吸氮量。

2.3.2 Slag Foaming

2.3.2 炉渣的发泡

2.3.2.1 Foaming Behavior of Slag

2.3.2.1 炉渣的发泡行为

Foaming is the phenomenon that gas is isolated from thin liquid film in liquid and can not move freely. The foaming of slag in metallurgical process can be understood as follows: There are a lot of bubbles in the slag liquid, which exist in the form of ball or polyhedron, and form a certain separation slag film between bubbles. The foaming process increases the surface energy, and the system reduces the surface area and eliminates the foam trend. However, because of the interface phenomenon between bubbles and separated slag film, the foam can exist in a certain time. The smaller the bubble diameter is, the smaller the stress on the bubble is, and the longer the bubble stays in the slag. And the larger the friction coefficient of the bubble moving in the slag, the larger the foaming range is.

泡沫化是气体在液体中被薄液体膜所隔离，不能自由运动的现象。冶金过程中熔渣的泡沫化可以理解为：炉渣液体中存在大量气泡，以球状或多面体状形式存在，并在气泡间形成一定的隔离渣膜；发泡过程增加了表面能，体系有减少表面积消除泡沫的趋势；由于气泡和分隔渣膜的界面现象，使泡沫能在一定时间内存在，气泡的直径越小，气泡受力越小，气泡在渣中滞留时间越长；气泡在熔渣中运动的摩阻系数越大，则发泡幅度就越大。

Effect of Surface Tension on Bubble Size in Slag
表面张力对熔渣中气泡尺寸的影响

The forming process of bubbles includes nucleation, growth, coalescence of small bubbles, floatation and elimination. The size of bubbles in slag is determined by these steps. The speed and radius of bubbles formed in the slag are directly proportional to the surface tension of the slag. When the surface tension is small, the nucleation speed of bubbles is large and the bubble radius is small. The surface tension has a great influences on the coalescence and growth rate of small bubbles in slag. The surface tension of slag and the discharge velocity of bubbles are large when they merge, so the merging growth velocity is large, too. On the contrary, the bubble coalescence growth rate of low surface tension system is small. The longer the retention time is in the slag. From the analysis of the nucleation and growth of bubbles in slag and the combination of bubbles, it can be concluded that if the surface tension of slag is small, the size of bubbles in slag will be small.

气泡的形成要经历形核、长大、小气泡的合并及上浮排除等过程。渣中气泡尺寸大小由这几个环节所决定：在熔渣中形成气泡的速度及气泡半径大小分别与熔渣表面张力成正比。表面张力小，则气泡的形核速度大，且气泡半径小。表面张力大小对熔渣中小气泡的合并、长大速度有很大影响。熔渣的表面张力大，气泡合并时的排液速度则大，因而合并长大速度就大。反之，则低表面张力体系的气泡合并长大速度就小，在渣中滞留时间就越长。由熔渣中气泡的形核长大及气泡的合并几个环节的分析可得：若熔渣的表面张力小，则渣中气泡的尺寸就小。

Effect of Surface Tension on Foaming Range
表面张力对发泡幅度的影响

When the surface tension of slag is small, the size of bubbles in foam slag is small and the distribution is uniform. At this time, the slag has a large foaming range. The surface tension decreased and the foaming range increased. However, when the surface tension decreases to a certain value, the foaming range decreases slightly. It can be seen that the slag with lower surface tension has higher foaming range. However, if it is too low, the elastic film will be lost, which will make the bubble brittle, rupture and reduce the foaming range.

渣的表面张力小，则泡沫渣中的气泡尺寸小，且分布均匀，此时熔渣具有较大的发泡幅度。表面张力降低，则发泡幅度增大。但当表面张力降至一定值以后，发泡幅度反而略有下降。由此可见较低的表面张力熔渣有较高的发泡幅度；但过低又会因失去弹性薄膜，使气泡变脆、破裂，从而降低发泡幅度。

2.3.2.2　Factors Affecting Slag Foaming

2.3.2.2　影响熔渣泡沫化的因素

The main influencing factors of slag include：

影响熔渣的主要影响因素包括：

（1）Surface tension：The surfactant in slag promoted the foaming of slag, and slag with lower surface tension was beneficial to the generation and maintenance of foam. According to $\Delta G = \Delta A \cdot \sigma$, slag foaming causes an increase in the interface between slag and liquid bubbles and increases the surface energy accordingly. The slag is in an unstable state, and the foam has a tendency to gradually eliminate. If the surface tension of the slag decreases, the amount of work needed to form the foam slag will be beneficial to the formation of slag foams.

（1）表面张力：渣中表面活性物质对于渣的泡沫化具有促进作用，表面张力较低的熔渣有利于泡沫的产生和维持。根据$\Delta G = \Delta A \cdot \sigma$，炉渣泡沫化造成体系渣液和气泡之间的界面增加，相应地使表面能增加，泡沫渣处于不稳定状态，其泡沫有逐渐消除的趋势；如果熔渣的表面张力减小，则形成泡沫渣所需要的功减少，有利于熔渣泡沫的形成。

（2）Viscosity：The foam of molten slag is formed by separating bubbles from molten slag liquid membrane. The existence of liquid film between bubbles after formation is an important factor affecting the existence of foam. In fact, the elimination of bubbles can also be regarded as the removal process of slag liquid, and the viscosity of slag is an important factor affecting the flow of slag. Proper increase of viscosity is beneficial to the existence of liquid film between bubbles, and it can also promote the formation and maintenance of slag.

（2）黏度：熔渣的泡沫是由熔渣液膜隔离气泡而形成的，气泡形成后它们之间的液膜能否持久存在是泡沫存在的重要影响因素。气泡的消除实质上也可以看成是气泡间渣液的排除过程，而熔渣黏度是影响渣液流动的重要因素。适当增加黏度有利于气泡间液膜的存在，也能促进泡沫渣的形成和维持。

（3）Density：If the density of the slag is larger, it is equivalent to adding a greater load on the foam structure, which will lead to the rupture of bubbles in the slag and reduce the foaming height.

（3）密度：如果熔渣的密度较大，这就相当于在泡沫结构上加上了更大的负载，这样会促使渣中气泡破裂，从而造成发泡高度降低。

（4）Influences of slag composition：The slag used for the treatment of synthetic slag is mainly $CaO-Al_2O_3-SiO_2-MgO-CaF_2$ slag system. The change of the relative content of each component in the slag causes the change of slag physical properties, resulting in the change of slag foaming properties.

(4) 炉渣成分的影响：合成渣处理所用的炉渣主要是 $CaO-Al_2O_3-SiO_2-MgO-CaF_2$ 渣系，渣中各组分相对含量的变化会相应引起炉渣物理性质的变化，从而引起炉渣泡沫化性能的改变。

(5) Alkalinity R: The results show that the foaming index of $CaO-SiO_2$ slag varies with alkalinity in wavy ways. When $R=1.22$, the foaming index is the lowest; when $R>1.22$, the foaming index increases with R; and when $R=1.9 \sim 2.0$, the foaming index is the highest. The increase of R results in the increase of $2CaO \cdot SiO_2$ in the slag, which increases the viscosity of the slag, hinders the movement of bubbles and prolongs the retention time of bubbles in the slag. In addition, the adhesion of solid particles to the Gas-Liquid interface also improves the strength and elasticity of the liquid membrane, which makes it difficult for the liquid membrane to break. The foaming index decreased after $R>2$.

(5) 碱度 R：研究表明，$CaO-SiO_2$ 系熔渣的发泡指数随碱度的增减呈波状变化。当 $R=1.22$ 时发泡指数最低；当 $R>1.22$ 时，发泡指数随 R 的增大而增大，并在 $R=1.9 \sim 2.0$ 时出现最高值。R 增大会引起熔渣中固相质点 $2CaO \cdot SiO_2$ 不断增多，从而提高了熔渣黏度，对气泡运动起到阻碍作用，延长了气泡在熔渣中的滞留时间。另外，固相质点黏附在气液界面上还会提高液膜的强度和弹性，从而使液膜难以破裂。在 $R>2$ 后发泡指数有下降的趋势。

(6) SiO_2: SiO_2 is a surface active substance, and increasing its content appropriately is good for foaming slag. But P_2O_5 and S are also surface active substances. And it is indicated that they are not good for foam production. Therefore, the surfactant of slag should not be generalized.

(6) SiO_2：SiO_2 属于表面活性物质，适量增加其含量有利于泡沫渣。而 P_2O_5 和 S 同样也是表面活性物质，有研究结果表明它们不利于泡沫的产生。所以，对于熔渣的表面活性物质也不应一概而论。

(7) Al_2O_3: It is indicated that the influence of Al_2O_3 on foamed slag is related to the content of fluorite in slag. When fluorite content is 5% (by mass), Al_2O_3 has a promoting effect on foaming. When fluorite content is 10% (by mass), Al_2O_3 has inhibitory effect on foaming.

(7) Al_2O_3：有研究表明，Al_2O_3 对泡沫渣的影响与渣中萤石含量有关。当萤石含量（质量分数）为 5%时，Al_2O_3 对泡沫化有促进作用；当萤石含量（质量分数）为 10%时，Al_2O_3 对泡沫化有抑制作用。

(8) Fluorite (CaF_2): Some researchers believe that fluorite in refining slag can reduce the surface tension of slag and promote the formation of foam. However, some people believe that fluorite in slag is not conducive to the foaming of slag, only when refractory particles exist in slag can foams. The results of this study are inconsistent, and also in FeO, MgO and other components.

(8) 萤石（CaF_2）：有的研究者认为精炼渣中的萤石可使熔渣的表面张力下降，促进泡沫的生成；但也有人认为渣中萤石不利于熔渣的泡沫化，只有渣中存在难熔的颗粒才会促进泡沫产生。这种研究结果的不一致，还表现在 FeO、MgO 等成分上。

The influence of slag composition on the foaming properties of slag is rather complicated, due to the complexity of slag composition affecting the physical properties of slag. For the study of slag foaming, due to the different metallurgical processes aimed at many researchers, the composition of the slag is quite different, and the research conditions (crucible material, experimental temperature, etc.) are not the same. So the results lack the basis for direct comparison. Even for the slag with the same composition range, some of the results obtained by different workers are quite different, and even have the opposite conclusion. It is necessary to study the foaming performance of LF refining slag by referring to the previous research results, but the research results of different experimental conditions and compositions of slag can not be directly applied.

炉渣成分对熔渣泡沫化性能的影响相当复杂，这是炉渣成分对熔渣物理性质影响的复杂性造成的。对于熔渣泡沫化的研究，由于许多研究者所针对的冶金过程不同，研究对象（熔渣）的成分差别很大，研究条件（坩埚材料、实验温度等）也不尽相同，所以其结果缺乏进行直接比较的依据。即使是针对相同成分范围的熔渣，不同的工作者所得到的结果也相差很大，甚至有相反的结论。在 LF 研究精炼渣的发泡性能时，借鉴前人的研究成果是必要的，但不能将不同实验条件、不同成分组成炉渣的研究结果直接套用。

2.3.2.3 Relationship between Foaming Properties and Content of Each Component

2.3.2.3 泡沫化性能与各成分含量之间的关系

CaO and SiO_2

氧化钙与二氧化硅

In order to ensure a good refining effect, it is required that there is a high amount of free CaO in the refining slag. However, when the amount of CaO is too large, the melting point of the slag increases significantly, CaO appears supersaturation, and the activity decreases significantly. So CaO in LF refining slag should be appropriate. Both the basicity of the refining slag and the influence of CaO on other properties of the slag should be properly considered. The increase of SiO_2 is disadvantageous to deoxidation and desulfurization.

为保证良好的精炼效果，要求精炼渣中含有较高的 CaO 量。当 CaO 量过大时，熔渣的熔点明显增加，CaO 出现过饱和，其活度显著下降。所以 LF 精炼渣中的 CaO 应该适量，既要保证精炼渣的碱度，又要适当考虑 CaO 对熔渣其他性质的影响。SiO_2 的增加对脱氧、脱硫不利。

The experimental results show that when the content of SiO_2 increases from 5% to 20% (by mass), the trend of slag foaming index is more obvious. The main reason is that SiO_2 is a surfactant, and the increase of its volume is favorable for reducing the surface tension of slag, and it promotes foaming of slag and maintaining foam.

实验结果表明，SiO_2 含量（质量分数）在 5%~20% 范围内增加时，渣发泡指数上升的趋势较为明显。这是因为 SiO_2 属表面活性物质，其量增加有利于熔渣的表面张力降低，促进熔渣发泡以及泡沫的维持。

For the CaO-SiO_2 slag system, the foaming property of slag is related to the ratio of CaO/SiO_2 in the slag composition. When the CaO/SiO_2 ratio is close to 2, the foaming index of the slag reaches the maximum value. The main reason is that there is a high melting point compound $2CaO \cdot SiO_2$ in the range of the experimental slag composition. There will be a lot of slag in the slag composition. $2CaO \cdot SiO_2$ solid particles are dispersed in molten slag. When the slag is foamed, these particles are distributed on the bubble wall, which increases the apparent viscosity of slag, and they slows down the velocity of slag discharge and maintains the existence of slag foam.

对于 CaO-SiO$_2$ 渣系，熔渣的泡沫化性能与熔渣成分中 CaO/SiO$_2$ 比值有关，CaO/SiO$_2$ 比值在接近于 2 时渣的发泡指数达到最大值。这是因为实验渣成分范围中存在高熔点化合物 2CaO·SiO$_2$；在此组成的熔渣中会有大量 2CaO·SiO$_2$ 固体微粒子弥散在熔渣中；当熔渣泡沫化以后，这些微粒分布在泡沫的气泡壁上，增加熔渣的表观黏度，减慢熔渣的排液速度，从而保持熔渣泡沫的存在。

In the $CaO-Al_2O_3-SiO_2$ slag system with high Al_2O_3 content, there are $2CaO-Al_2O_3-SiO_2$ components with high melting point in the composition range of experimental slag. If the content of Al_2O_3 in LF refining slag is kept high, and the content of SiO_2 in LF refining slag changes gradually from small to large, the composition point of refining slag will pass near the component point of $2CaO-Al_2O_3-SiO_2$. If the content of Al_2O_3 in LF refining slag changes little, the component of $2CaO-Al_2O_3-SiO_2$ in ternary system has the same effect as that in binary system. The proper addition of SiO_2 in LF refining slag can improve the foaming performance of slag. But if its content is too high, the refining performance of refining slag will inevitably be affected.

在 Al$_2$O$_3$ 含量比较高的 CaO-Al$_2$O$_3$-SiO$_2$ 渣系中，实验渣成分范围内存在熔点较高的 2CaO-Al$_2$O$_3$-SiO$_2$ 组元。如果保持 LF 精炼渣中较高的 Al$_2$O$_3$ 含量，并由小到大逐渐变化渣中 SiO$_2$ 的含量，则精炼渣的成分组成点会通过 2CaO-Al$_2$O-SiO$_2$ 组元点附近；如果精炼渣中 Al$_2$O$_3$ 含量变化不大，那么三元系中 2CaO-Al$_2$O$_3$-SiO$_2$ 组元具有与二元系中 2CaO-SiO$_2$ 类似的作用。LF 精炼渣中 SiO$_2$ 成分的适量增加能够提高熔渣的泡沫化性能，但如果其含量过高，势必影响精炼渣的精炼性能。

CaF$_2$

氟化钙

CaF_2 has two effects on foaming effect: One is the increase of its amount and the decrease of surface tension, which is conducive to slag foaming; The other is the increase of its amount makes slag viscosity decrease, which is not conducive to foaming.

CaF$_2$ 对发泡效果影响有两方面：一是增加 CaF$_2$ 量，降低表面张力，有利于炉渣发泡；二是增加 CaF$_2$ 量，使炉渣黏度降低，其不利于炉渣发泡。

The experimental results show that the foaming index of slag varies greatly due to the content of CaF_2. Proper CaF_2 content can enhance the foaming properties of slag. The experiment shows that the dosage of CaF_2 is less than 5% (by mass).

实验结果显示，熔渣发泡指数因为 CaF$_2$ 含量而有很大变化，适当的 CaF$_2$ 含量可以增强渣的泡沫化性能。该实验表明 CaF$_2$ 的用量要在 5%（质量分数）以下。

2.3.3 Foaming Agent for LF Submerged Arc Refining Slag

2.3.3 LF 埋弧精炼渣的发泡剂

The gas sources needed to produce foamed slag are different in different processes. The gas source in the slag of EAF can come from the reaction between the carbon powder and the slag. The source of the emulsified gas in the oxygen top blown converter is mainly the reaction of carbon and oxygen. In the ladle refining process, there is not enough reaction for producing gas, and additional foaming agent is needed to provide enough gas source for the foaming of the refining slag.

在不同的工艺过程中，产生泡沫渣所需要的气源不同，电弧炉泡沫渣中的气源来自碳粉与炉渣之间的反应；氧气顶吹转炉中乳化的气源来自碳氧反应；而在钢包精炼过程中没有足够产生气体的反应，因此需要外加发泡剂，为精炼渣的泡沫化提供足够气源。

The higher temperature and reducing atmosphere conditions in LF refining furnace are not conducive to foaming of slag, and the foaming of slag is difficult to maintain for a longer time. In the experiment, it is found that the foaming height of the slag will be greatly reduced in the short time after stopping blowing. If there is no good gas source, even if the refining slag with good foaming performance can not form ideal foam slag, it is necessary to add foaming agent into the refining furnace.

LF 精炼炉中的较高温度和还原性气氛条件不利于炉渣的发泡，熔渣的泡沫也很难维持较长的时间。实验中发现基础渣吹气发泡后，在停止吹气的很短时间内，熔渣的泡沫高度会大大降低。如果没有好的气体源，即使是泡沫化性能良好的精炼渣也不可能形成理想的泡沫渣，因此精炼炉内加入发泡剂是必要的。

2.3.3.1 Selection of Foaming Agent

2.3.3.1 发泡剂的选择

In order to make the slag foam, it is necessary to ensure that the basic slag has proper physical properties, such as larger viscosity, smaller surface tension and suitable alkalinity. In addition, there should be enough gas source. The electrode reacts with the slag. Argon agitation can provide parts of the gas source, and the gas can also be produced by adding foaming agent. The selected foaming agent has the following characteristics:

要使炉渣泡沫化，就要保证精炼基础渣有适宜的物理性质，即较大的黏度，较小的表面张力，以及适宜的碱度。另外还要有足够的气源，电极与炉渣反应，氩气搅拌提供一部分气源，也可通过外加发泡剂产生气体。选用的发泡剂应该具备以下特点：

(1) Under the condition of LF production, it can provide the gas needed for foaming and supply continuously for a certain period of time.

(1) 在 LF 生产条件下，可以提供泡沫化所需的气体，并在一定时间内持续供给。

(2) The foaming agent selected can not bring great harm to the ladle refining function.

(2) 所选用的发泡剂不能对钢包精炼的功能带来大的危害。

(3) It has convenient application and low price.

(3) 应用方便，价格低廉。

The materials suitable for foaming agent can be divided into three categories: carbonate, chloride and fluoride.

适合作发泡剂的材料可分为 3 类,它们主要是碳酸盐、氯化物和氟化物。

Carbonate

碳酸盐

Limestone, dolomite and industrial alkali are commonly used. Under high temperature, the following reactions are as follows:

常用的碳酸盐有石灰石、白云石和工业碱,在高温下主要发生以下反应:

$$CaCO_3 = CaO+CO_2 \tag{2-5}$$
$$MgCO_3 = MgO+CO_2 \tag{2-6}$$
$$BaCO_3 = BaO+CO_2 \tag{2-7}$$

It is easy to decompose and produce gas products at high temperature, and the decomposition products of some carbonates are just the components of refining slag. Calcium carbonate and magnesium carbonate are relatively easy to obtain candidate materials. Several kinds of carbonates can obtain different degrees of foaming effect. Their common characteristics are that after the foaming agent is added to the base slag, the height of the slag foam reaches a maximum value due to the decomposition of carbonate, and the volume of gas decreases and the slag height decreases due to the consumption of foaming agent. In a certain period of time, due to the delay of micro bubbles in slag, the foaming effect is kept moderate. The peak value of $MgCO_3$ is higher than that of $CaCO_3$ and $BaCO_3$ at the same mass. The main reason is that the molecular weight of $MgCO_3$ is small, and the moles of $MgCO_3$ at the same mass are larger, which can provide more decomposition gas. However, the peak value of foam disappears quickly. The foaming effect of $BaCO_3$ as foaming agent is obviously lower than that of $MgCO_3$ and $CaCO_3$, but the change of slag height is relatively gentle. When $CaCO_3$ is used as foaming agent, the peak value of foaming is higher, and the height of the slag after the peak is lower than that of $MgCO_3$, and the foam holding time is longer. In addition, CaO produced by the decomposition of $CaCO_3$ also maintains the basicity of slag, which is also beneficial to the refining effect of slag. Therefore, $CaCO_3$ is suitable as the basic material of foaming agent for LF refining slag.

在高温下容易分解并产生气体产物,而且有些碳酸盐的分解产物正好是精炼渣的组成部分,碳酸钙、碳酸镁等是比较容易得到的候选材料。几种碳酸盐都可以得到不同程度的发泡效果,它们的共同特征是在发泡剂加入基础渣后的较短时间内,由于碳酸盐的分解使得基础渣泡沫高度达到一个极大值,然后由于发泡剂的消耗造成气量减少,渣高度下降。在一定的时间内由于渣中微小气泡排出的延迟而保持适度的发泡效果。在相同质量的用量情况下,$MgCO_3$ 相对于 $CaCO_3$ 和 $BaCO_3$,其发泡峰值较高。主要是因为 $MgCO_3$ 的分子量较小,在相同质量下 $MgCO_3$ 的摩尔数较大,可以提供更多的分解气体,但到达其峰值后泡沫的消失速度很快。$BaCO_3$ 作为发泡剂,发泡效果明显不如 $MgCO_3$ 和 $CaCO_3$,但其渣高度的变化比较平缓。$CaCO_3$ 作为发泡剂时的发泡峰值也较高,而且达到峰值后渣泡沫高度的降低比 $MgCO_3$ 慢,泡沫维持时间相对较长。另外,$CaCO_3$ 分解所产生的 CaO 同时也维持了渣的碱度,对渣的精炼作用也有好处。因此作为 LF 精炼渣发泡剂的基础材料,$CaCO_3$ 是较为合适的。

Carbon Powder and Carbide
碳粉和碳化物

Carbon powder and carbide are commonly used slag foaming agents in electric furnace production. The main foaming principle is that they can react with the oxidizing substances in the slag to produce gas, but due to the characteristics of the ladle refining furnace, the oxidation of the refining slag is very low. So if carbon powder and carbide are used as foaming agent alone in the ladle refining furnace, the effect will be very different from the traditional electric furnace.

碳粉和碳化物是在电炉生产中常用的炉渣发泡剂，其主要的发泡原理是能够与炉渣中的氧化物质进行反应产生气体。但由于钢包精炼炉的特点，精炼渣的氧化性非常低。因此如果碳粉和碳化物单独作为发泡剂使用在钢包精炼炉中的效果与传统的电炉会有很大的区别。

The common are coke, silicon carbide and carbide. Since the oxygen and FeO in the steel are high at the beginning of LF furnace, these substances will react with the slag:

常见的碳粉和碳化物有焦炭、碳化硅和电石。由于 LF 炉开始阶段钢中氧和渣中 (FeO) 均较高，这些物质与炉渣主要发生以下反应：

$$C+(FeO) = [Fe]+CO \tag{2-8}$$

$$SiC+3(FeO) = 3[Fe]+(SiO_2)+CO \tag{2-9}$$

$$CaC_2+3(FeO) = 3[Fe]+(CaO)+2CO \tag{2-10}$$

Compound Foaming Agent of Carbonate and Carbide
碳酸盐与碳化物复合发泡剂

The foaming effect of carbonates and carbides is improved. The foaming effect of foaming agent is prolonged, which is beneficial to the retention of base slag foam. The main reason is that the CO_2 produced after the decomposition of carbonate reacts with the carbide contained in the compound foaming agent at high temperature to form CO, which increases the air volume with more favorable for the foaming of the base slag. The decomposition rate of carbonate at high temperature is fast, the reaction time is short, and the volume of gas produced is less. The results show that the foaming effect of foaming agent mainly composed of SiC and CaC_2 is better. But the foaming effect of CaC_2 is better than that of CaC_2. Generally, the foaming agent of mixed SiC and CaC_2 has the best foaming effect. It is worth noting that although CaC_2 has good foaming effect, it is difficult to transport and preserve. The foaming effect of SiC and CaC_2 foaming agents is significant when the content of iron oxide in slag is high (in the early stage of LF electroslag), but the foaming ability will be limited when the content of oxygen in steel and slag is low (in the later stage of LF), that is to say, in the later stage of deoxidization. For low silicon steel, it is necessary to pay attention to the problem of increasing silicon in molten steel caused by the reduction of SiC in foaming agent and SiO_2 in slag.

碳酸盐与碳化物等组成复合发泡剂改善碳酸盐的发泡效果。其表现在发泡剂的发泡效果时间得到延长，这有利于基础渣泡沫的保持。主要是因为碳酸盐分解后产生的 CO_2 与复合发泡剂中所含的碳化物在高温下进行部分反应生成 CO，从而增加了气量，对基础渣的发泡更有利。碳酸盐在高温下的分解速度快，反应时间短，且产生的气体体积也较少，因此采用以 SiC 和 CaC$_2$ 为主的发泡剂发泡效果较好，CaC$_2$ 发泡效果更好。通常 SiC 和 CaC$_2$ 混合型的发泡剂具有最好的发泡效果。尽管 CaC$_2$ 具有良好的发泡效果，但运输和保存比较困难。SiC 和 CaC$_2$ 型的发泡剂在渣中氧化铁含量较高时（LF 通电造渣前期）、发泡效果显著，而当钢、渣中氧含量较低（LF 后期），即到脱氧后期时，其发泡能力将受到明显限制。对于低硅钢还要注意发泡剂中 SiC 及渣中 SiO$_2$ 被还原造成的钢水增硅问题。

CaCl$_2$

氯化钙

It has a good foaming effect, but when it is used in mass production, CaCl$_2$ will have the following shortcomings:

CaCl$_2$ 具有较好的发泡效果，但当其大批量用于生产时，CaCl$_2$ 会存在以下不足：

(1) The water absorption is strong. If the storage time is long, the water content of the product will exceed the standard and affect its foaming effect.

(1) 吸水性强。如果储存时间长，会使产品水分超标，并影响其发泡效果。

(2) High temperature decomposition of CaCl$_2$ will produce Cl$_2$, which is harmful to the environment.

(2) CaCl$_2$ 高温分解会产生 Cl$_2$，对环境有害。

(3) The cost is high.

(3) 成本较高。

2.3.3.2　Effect of Foaming Agent Size and Adding Method

2.3.3.2　发泡剂粒度及加入方式的影响

In the foaming process of calcium carbonate with different particle size in slag, the foaming agent should have a certain particle size to keep the foaming effect for a long time. The mixed use of different granulant blowing agents can improve the reaction of small particle foaming agent, and the foam is not easy to maintain. It can also enhance the initial foaming strength of large size foaming agent.

不同颗粒大小的碳酸钙在渣中的发泡过程中，发泡剂应具有一定粒度，从而使发泡效果得到较长时间保持。不同粒度发泡剂的混合使用可以改善小粒度发泡剂反应猛烈、泡沫不易保持的缺点，也可以使大粒度发泡剂的初期发泡强度有所增强。

2.3.3.3　Amount of Foaming Agent

2.3.3.3　发泡剂用量

The amount of slag foaming agent is an important factor affecting slag foaming. In a certain range, increasing the amount of foaming agent can effectively improve the foaming height of slag. However, when the physical and chemical properties, slag layer thickness and temperature are fixed, and the ability of slag to store gas is basically stable. When the amount of gas produced by foaming agent exceeds the maximum storage capacity of slag, it is impossible to increase the foaming height of slag with further addition of foaming agent. Using foam agent to carry out submerged arc heating of foam slag, the reasonable addition amount of foaming agent should be determined according to the actual situation of slag.

炉渣发泡剂的加入量是影响炉渣发泡的重要因素。在一定范围内,增加发泡剂的用量可以有效提高炉渣的发泡高度。但由于炉渣的理化性能、渣层厚度和温度一定,炉渣储存气体的能力也基本稳定。当发泡剂产生的气体量超过了炉渣的最大储存能力时,再增加发泡剂加入量,也不可能再提高炉渣的发泡高度。在使用发泡剂进行泡沫渣埋弧加热时,应根据炉渣的实际情况,确定发泡剂的合理加入量。

2.3.4 Impact of Operation Process

2.3.4 操作工艺的影响

The operation conditions also have important influences on the foaming process of LF, which are as follows:

操作工艺条件对于 LF 生产过程中的泡沫渣工艺也有重要影响,比如:

(1) Refining temperature: Temperature rise is not conducive to the stability of foam. Because of the influence of temperature on the surface tension and viscosity of slag, it is necessary to pay attention to the control of treatment temperature in the production of ladle refining furnace.

(1) 精炼温度:温度升高不利于泡沫的稳定。由于温度对熔渣的表面张力和黏度都有一定的影响,因此钢包精炼炉的生产中要注意温度的控制。

(2) The way of refining slag: The mixture of foaming agent and refining slag is beneficial to the high efficiency of foaming agent, and the effect is similar to that of internal gas source.

(2) 精炼渣:发泡剂与精炼渣混合加入炉中有利于发泡剂高效地发挥作用,取得近似于内部气源的效果。

(3) The way of foaming agent: Batch addition of foaming agent is beneficial to the sustained maintenance of foam source in the furnace, and a longer foam effect can be obtained.

(3) 发泡剂:发泡剂分批添加有利于炉中泡沫气源的持续维持,可以得到较长的泡沫效果。

2.3.4.1 Determination of Slag Thickness and Slag Material

2.3.4.1 渣厚及渣料确定

The stable submerged arc slag thickness is greater than 80mm (due to the arc length of 70~80mm), the slag discharge of converter is 30~50mm, and the new slag material thickness is at least 30~50mm. Taking into account the technological characteristics of foam slag, the slag should be properly reduced. At the same time, calcium carbide should be added in order to ensure the processing time of 20~30min to maintain foam residue. It is believed in the literature that the thickness of slag layer must be more than 2 times of arc length for submerged arc heating.

稳定的埋弧渣厚应大于 80mm(因弧长为 70~80mm)。因转炉下渣量为 30~50mm,要求新加渣料厚度至少 30~50mm。考虑到泡沫渣工艺特点，渣料应适当减少。同时为了确保处理时间维持泡沫渣保持在 20~30min 之间，还应加入电石。埋弧加热要求炉渣的渣层厚度必须达到电弧长度的 2 倍以上。

The arc length L is related to the phase voltage U, phase current I, phase resistance R and power factor $\cos\varphi$. The equation is given by：

电弧长度 L 与相电压 U、相电流 I、电阻 R 和功率因数 $\cos\varphi$ 有关，其计算公式为：

$$L = U\cos\varphi - IR - 30 \tag{2-11}$$

When the arc voltage U_h is known, the arc length can be simply calculated by：

当已知电弧电压 U_h 时，电弧长度的计算公式为：

$$L = U_h - 30 \tag{2-12}$$

2.3.4.2 Machine Mixed Slag and Premelted Slag

2.3.4.2 机混渣与预熔渣

Mechanical Mixed Slag

机械混合渣

Due to the difference of component proportion, mechanical mixed slag is easy to produce component segregation; its performance is unstable; it is easy to hydrate when it is placed for a long time, which affects the refining effect; and the rate of slag formation and desulfurization is lower than that of premelted slag.

由于组分比重的差异，机械混合渣易产生成分偏析；其性能不稳定；放置时间长时易水化，影响精炼效果；成渣和脱硫速度均低于预熔渣。

Premelted Slag

预熔渣

Premelted slag is stable composition, less moisture, not easy to absorb moisture during transportation and storage. The following effects can be achieved in actual use：

预熔渣成分稳定，水分少，运输和保存过程中不易吸潮。实际使用中可达到以下效果：

（1）It can shorten the refining time, reduce the power consumption and electrode consumption.

（1）加快成渣速度，改善脱硫的动力学条件，可缩短精炼时间，从而降低电耗和电极消耗。

（2）Improve the effect of refining desulfurization, especially for deep desulfurization.

（2）提高精炼脱硫效果，特别是适合深脱硫。

（3）Improve the life of ladle lining.

（3）提高钢包精炼炉包衬寿命。

（4）Reduce hydrogen content in steel.

（4）减少钢中氢含量。

（5）Reduce the refining cost of LF furnace.

（5）降低 LF 炉精炼成本。

2.3.5　Detection and Control Technology of Foam Residue

2.3.5　泡沫渣的探测和控制技术

In order to ensure high quality and proper amount of foam residue in the electric furnace, CRM (Belgium Steel Research Center) has developed a technology that can automatically detect and control foamed slag generation automatically under the support of CECA (European Coal and Steel Union), CRM has installed two kinds of sensors on the 100t LF refining furnace of Gustabule Factory of Hogovin Group in the region of Laluvier. One is the vibration frequency meter, which is used to record the sound released by the furnace; The other is the accelerometer, which is used to measure the vibration of the furnace body, as shown in Figure 2-3.

为了保证电炉内产生高质适量的泡沫渣，在 CECA（欧洲煤钢联合体）的资助下，CRM（比利时的钢铁研究中心）开发研究出一项能自动动态探测和控制泡沫渣生成的技术。CRM 在比利时拉卢维耶尔地区的霍戈文集团古斯塔布勒厂的 100t LF 精炼炉上安装了两种传感器，一是振动频率计，用于记录炉子释放的声响；另一种是加速计，用于测量炉体的振动，如图 2-3 所示。

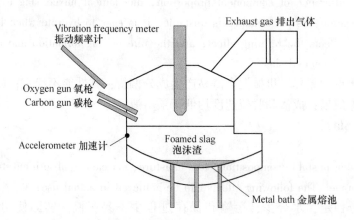

Figure 2-3　Installation and application of two kinds of sensors in LF refining furnace
图 2-3　两种传感器在 LF 精炼炉上的安装及应用

The vibration frequency meter is installed in the control room, directly facing the furnace door, and connected with the PC and fluorescent screen of the control room through a pre amplifier, a two stage selection filter and an amplifier. The accelerometer is located on the side wall of the furnace cooling wall and directly contacts with the foam slag. The sensor is embedded in the heat resistant resin box and placed in the furnace. The sensor is connected to the PC in the same way as the vibration frequency meter. The frequency of the signal collected by the sensor is analyzed by the fourier transform method. And then its formation characteristics are reflected. During the production process, the system tracking found that the signal amplitude of the sensor is small, and the foam quality and volume in the furnace are suitable.

振动频率计安装在控制室内,直对炉门,经过一个前置放大器、一个两级选择过滤器和一个放大器与控制室的 PC 机及荧光屏连接。加速计位于炉体侧边水冷却壁上,其直接与泡沫渣接触;传感器被嵌入耐热的树脂盒中,放在炉内。该传感器和振动频率计一样,以同样的方法与 PC 机相连。通过傅立叶变换法对传感器收集到的信号进行频率分析,继而反映出它的形成特征。在生产过程中系统跟踪发现,传感器的信号振幅小,炉内的泡沫质量和体积适宜。

Task 2.4 Typical Steel LF Refining Process
任务 2.4 典型钢种 LF 精炼工艺

In addition to ultra-low carbon, nitrogen, sulfur and other ultra-pure steels, almost all kinds of steel can be refined by LF method, especially for bearing steel, alloy structural steel, tool steel and spring steel. After refining, the total oxygen content of bearing steel is reduced to 0.001% (by mass), [H] to 0.0003%~0.0005% (by mass), [N] to 0.0015%~0.002% (by mass), and the total amount of nonmetallic inclusions is 0.004%~0.005% (by mass).

除超低碳、氮、硫等超纯钢外,几乎所有的钢种都可以采用 LF 法精炼,特别适合轴承钢、合金结构钢、工具钢及弹簧钢等的精炼。精炼后轴承钢全氧含量降至 0.001%(质量分数),[H] 降至 0.0003%~0.0005%(质量分数),[N] 降至 0.0015%~0.002%(质量分数),非金属夹杂物总量在 0.004%~0.005%(质量分数)。

2.4.1 General Carbon and Low Alloy Refining Process
2.4.1 普碳和低合金精炼工艺

For ordinary carbon steel and general low alloy steel, it is very difficult to reduce the total oxygen in the steel to below $30mg/m^3$ by means of diffusion deoxidization in LF station due to the fast production speed (LF treatment time is less than 25min). Therefore, desulfurization and composition adjustment are the main tasks of refining.

由于生产节奏快(LF 处理时间小于 25min),普碳钢和一般的低合金钢在 LF 工位依靠扩散脱氧方式把钢中全氧降到 $30mg/m^3$ 以下非常困难。因而精炼以脱硫和调整成分为主要任务。

In the process of LD tapping, it is necessary to increase the amount alloy of Si, Al and Ba, to stabilize the acid soluble aluminum of LF incoming steel, and to strengthen the precipitation deoxidization of steel. The LF station adjusts the composition of the top slag, absorbs the deoxidation products floating in the steel water, inhibits the oxygen transfer from the top slag to the molten steel, and advances the actual deoxidation task of the molten steel to the LD tapping process, thus reducing the LF refining load. The refining slag system uses slag system 1 in Table 2-4, which has high basicity and desulfurization rate of more than 80%; Generally speaking, for the molten steel without hot metal pretreatment, the sulfur content of converter tapping is more than 0.040%

(by mass). The sulfur content of steel can be reduced to less than 0.015% (by mass) and the total slag content can be controlled within 1.0%~1.5% (by mass) by LF refining in a short time.

在 LD 出钢过程中，需要增加硅、铝、钡合金用量，稳定 LF 进站钢水酸溶铝，强化钢水沉淀脱氧。LF 工位调整顶渣成分，吸附钢水中上浮的脱氧产物，抑制顶渣含氧向钢水传递，把钢水实际的脱氧任务提前到 LD 出钢过程中来完成，从而减轻 LF 精炼负荷。精炼渣系采用表 2-4 中渣系 1，该渣系碱度高，脱硫率在 80% 以上。对于没有铁水预处理的转炉钢水，一般转炉出钢硫在 0.040%（质量分数）以上。在较短的时间内，可以将钢中硫含量（质量分数）降到 0.015% 以下，总渣量控制在 1.0%~1.5%（质量分数）。

Table 2-4 LF refining slag system range
表 2-4 LF 精炼渣系范围

Slag system 渣系	Component (mass fraction)/% 化学成分（质量分数）/%				
	CaO	SiO_2	MgO	Al_2O_3	CaF_2
Slag system 1 渣系 1	50~55	10~15	6~10	—	20
Slag system 2 渣系 2	40~45	20~25	5~6	10~15	
Slag system 3 渣系 3	45~50	10~20	8~15	10~15	—

2.4.2 Refining Process of Medium and High Carbon Steel

2.4.2 中高碳钢精炼工艺

Metal products (such as medium and high carbon hard wire, bearing steel, spring steel, etc.) should not only reduce the total amount of inclusions in the steel, but also pay attention to control the particle size of alumina and aluminum containing composite inclusions and their distribution in the slab, so as to ensure the drawing performance and fatigue performance of the steel. There are higher requirements for inclusion size, type and gas content in steel. At present, there are two kinds of refining slag with high basicity ($CaO/SiO_2>3$) and low basicity ($CaO/SiO_2\approx2$).

金属制品用材（如中高碳硬线、轴承钢、弹簧钢等）不仅要降低钢中夹杂总量，更要注意控制氧化铝及含铝复合夹杂物的粒度及在铸坯中的分布，以保证钢材的拉拔性能和疲劳性能。对钢中夹杂物尺寸、类型、气体含量均有较高要求。目前中高碳精炼渣有高碱度（$CaO/SiO_2>3$）和低碱度（$CaO/SiO_2\approx2$）两类。

High basicity slag has the advantages of large sulfur capacity and low oxygen content in refined steel, but it is easy to form calcium aluminate type spherical inclusions. In addition, due to the poor fluidity of high basicity slag and improve the fluidity, it is necessary to increase the amount of fluorite added, which increases the foreign inclusions brought by furnace lining erosion.

高碱度渣具有硫容量大、精炼的钢含氧低的优点，但易生成铝酸钙型球状夹杂物。另外，由于高碱度渣流动性不好，为提高流动性需提高萤石的加入量，加大炉衬浸蚀带入的外来夹杂物。

The slag with low basicity is easy to absorb inclusions in steel, so the quantity of inclusions in steel is reduced, and its shape and properties can be effectively controlled, especially beneficial to eliminate the point inclusions in steel. Moreover, due to the low basicity, the fluidity of slag is good, which can promote the metallurgical reaction speed, and the number of sulfide inclusions can be controlled in a satisfactory range. However, the slag system with low basicity is required to be equipped with hot metal pre desulfurization to reduce the desulfurization pressure of LF furnace.

低碱度渣易于吸收钢中夹杂物，使得钢中夹杂物数量减少，其形态和性质也能得到有效的控制，对消除钢中点状夹杂物十分有利。由于碱度低，炉渣的流动性好，低碱度渣能促进冶金反应速度，同时硫化物夹杂的数量能控制在一定范围内。采用低碱度渣系，要求配有铁水预脱硫，以减轻LF炉的脱硫压力。

In the actual production, the refining slag system adopts the two components of the slag system in Table 2-4, the total slag volume is controlled at about 2% (by mass), the refining time is controlled at 40~45min, and the appropriate power supply system and argon blowing system are selected to avoid nitrogen increase in the refining process. The deoxidizing alloy adopts low aluminum or non aluminum alloy. After the refining, calcium treatment is adopted to modify the inclusions and ensure the weak mixing time is more than 8min. After using the slag system, the inclusion content in the steel is mostly controlled below 5mg/L, and the deformation performance is good, which effectively eliminates the drawing embrittlement caused by inclusions.

实际生产中精炼渣系采用表2-4中渣系2组分，总渣量控制在2%（质量分数）左右，精炼时间控制在40~45min，同时选用合适的供电制度及吹氩制度，避免精炼过程增氮。脱氧合金均采用低铝或无铝合金，精炼完毕采用钙处理，对夹杂物进行改性，并保证弱搅拌时间>8min。应用该渣系后，钢中夹杂物含量大部控制在5mg/L以下，而且变形性能良好，能有效杜绝因夹杂物引起的拉拔脆断。

2.4.3 Refining Process of High Quality Plate

2.4.3 优质板材精炼工艺

For pipeline steel, low-carbon steel and low sulfur high-quality plate steel, the refining slag system is the slag system 3 in Table 2-4. The molten iron into the furnace is pretreated and desulfurized first. The slag is modified by converter tapping. The LF position controls aluminum in the whole process, and carbide and other carbon containing slag materials are not used. The basicity of the slag system is 2.5~3.5, and the desulfurization effect is good. The amount of slag has great influence on deep desulfurization. In order to adjust the amount of slag, the proper amount of lime and synthetic slag should be added in the LF station. Meanwhile, the viscosity of slag should be considered. In the actual production, the slag of 2.0%~2.5% should be selected.

对于管线钢、低碳钢等低硫优质板材用钢，精炼渣系采用表2-4中渣系3。入炉铁水先进行预处理脱硫，转炉出钢对氧化渣进行改质，LF位全程控铝，并且不使用电石等含碳渣料。该渣系碱度为2.5~3.5，脱硫效果良好。渣量对深脱硫的影响比较大。转炉出钢时要加入适量石灰和合成渣，在LF站再加入适量造渣剂，以调整渣量，同时要考虑炉渣的黏度，实际生产中选择2.0%~2.5%（质量分数）的渣量。

2.4.4 Slag for Refining Low Carbon Aluminum Containing Steel

2.4.4 低碳含铝钢精炼用渣

$CaO-SiO_2$ slag system is difficult to be used in low–carbon aluminum containing steel because of its weak desulfurization ability and high acid soluble aluminum in low–carbon aluminum containing steel, which has reduction effect on SiO_2 in slag. More Al_2O_3 is produced by aluminum deoxidation, and the content of these products can reach more than 10% in refining slag. Therefore, the $CaO-CaF_2$ slag system is not likely to be used in the refining slag of low-carbon aluminum containing steel, but the $CaO-Al_2O_3$ slag system is more likely to be used.

由于CaO-SiO$_2$渣系脱硫能力较弱，并且低碳含铝钢中酸溶铝较高，对渣中（SiO$_2$）有还原作用，因此在低碳含铝钢上难于采用CaO-SiO$_2$渣系。由于铝脱氧生成的Al$_2$O$_3$较多，这些产物的含量（质量分数）在精炼渣中达到10%。因此，低碳含铝钢精炼渣也不太可能采用CaO-CaF$_2$渣系，而是倾向于采用CaO-Al$_2$O$_3$渣系。

In order to make the refining slag have better desulfurization effect, at the same time, facilitate the assimilation and absorption of deoxidized products floating, the composition of the final refining slag is often selected in the $12CaO \cdot 7Al_2O_3$ formation area of the $CaO-Al_2O_3-SiO_2$ phase diagram. When Al_2O_3 is about 30% (by mass) or $CaO/Al_2O_3 = 1.8$, there is an area with higher L_s, as shown in Figure 2-4.

为了使精炼渣具有较好的脱硫效果，以及有利于对上浮Al$_2$O$_3$等脱氧产物的同化和吸收，常将精炼终渣成分选定在CaO-Al$_2$O$_3$-SiO$_2$相图的12CaO·7Al$_2$O$_3$生成区域，该区域中，当Al$_2$O$_3$含量为30%（质量分数）或CaO/Al$_2$O$_3$=1.8时，L_s较高，如图2-4所示。

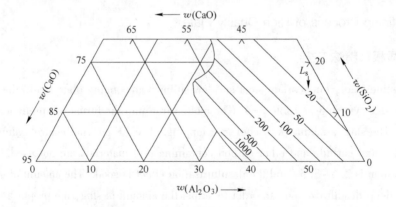

Figure 2-4 Sulfur distribution ratio of $CaO-Al_2O_3-SiO_2$ ternary slag system
[1600℃, $w[Al] = 300ppm$, $w(MgO) = 5\%$]

图2-4 CaO-Al$_2$O$_3$-SiO$_2$三元渣系硫分配比
[1600℃, $w[Al] = 300ppm$, $w(MgO) = 5\%$]

In this area, the melting point of refining slag is low, which is beneficial to the combination with inclusions. More importantly, the reoxidation of [Al] in steel water caused by SiO_2 can be inhibited under this slag system. Therefore, the design refining final slag composition is: $w(CaO) = 55\% \sim 60\%$, $w(SiO_2) = 4\% \sim 7\%$, $w(Al_2O_3) = 28\% \sim 32\%$, $w(CaO)/w(Al_2O_3) = 1.7 \sim 1.9$.

在该区域中，精炼渣熔点较低，有利于与夹杂物结合。更重要的是，在这种渣系条件下由 SiO_2 引起的钢水中 [Al] 的再氧化趋势能得到抑制。设计精炼终渣组成为：$w(CaO) = 55\% \sim 60\%$，$w(SiO_2) = 4\% \sim 7\%$，$w(Al_2O_3) = 28\% \sim 32\%$，$w(CaO)/w(Al_2O_3) = 1.7 \sim 1.9$。

2.4.5 Smelting of Bearing Steel

2.4.5 轴承钢的冶炼

The typical smelting process of bearing steel is electric furnace process, but the non-ferrous metals brought in by scrap steel will increase the content of harmful elements in the steel and affect the service life of bearing steel. In order to reduce the residual elements in the steel and the production cost, Sumitomo, Thyssen and Kawasaki have successively developed and used the process of converting molten iron with less impurities to produce bearing steel, and gradually cooperated with LF ladle refining and RH vacuum treatment process to produce bearing steel with high purity, such as 60t BOF→60t LF furnace→60t VD→continuous casting→rolling.

轴承钢的典型冶炼工艺是采用电炉工艺，但废钢中带入的有色金属会增加钢中有害元素的含量，影响轴承钢寿命。为了降低钢中残余元素和生产成本，日本住友、德国蒂森和日本川崎制铁公司先后开发采用了杂质少的铁水经转炉吹炼生产轴承钢的工艺，并逐步配合 LF 钢包精炼和 RH 真空处理工艺生产出高纯净度的轴承钢。例如，60t BOF→60t LF 炉→60t VD→连铸→轧钢。

Among them, argon blowing is an effective measure to ensure the reaction in the refining furnace, the floatation of inclusions and the reduction of oxygen content. Different argon blowing systems are used in different stages of refining. In order to speed up the reaction during LF refining, the gas supply is increased properly, and the argon pressure is controlled at $0.6 \sim 1.0$ MPa. After the refining, the liquid steel is weakly stirred, in order to ensure the inclusion floats up, and the argon gas volume isn't too large during the weakly stirred, so as to prevent the liquid steel from churning severely and contacting with the air, secondary oxidation of the treated liquid steel or rolling the slag into the liquid steel. Generally, the argon blowing pressure is controlled at $0.2 \sim 0.3$ MPa, and the weak blowing time is controlled at least 7 min.

其中，吹氩是保证精炼炉内反应，以及夹杂上浮、氧含量降低的有效措施。在精炼的不同阶段采用不同的吹氩制度。为了在 LF 精炼过程中加速反应进行，应适当加大供气量，氩气压力控制在 $0.6 \sim 1.0$ MPa；精炼结束后，为保证夹杂上浮，要对钢水进行弱搅拌，弱搅拌时氩气量不可过大，防止因钢水翻腾厉害与空气接触，将已经处理好的钢液二次氧化，或将渣子卷进钢液。一般吹氩压力控制在 $0.2 \sim 0.3$ MPa，弱吹时间控制在 7min 以上。

The $CaO-SiO_2-Al_2O_3$ slag system with high basicity is adopted for treatment, and its slag system composition is shown in Table 2-5.

在冶炼过程中采用高碱度 $CaO-SiO_2-Al_2O_3$ 渣系进行处理,其渣系组成见表 2-5。

Table 2-5 Slag washing for LF refining bearing steel

表 2-5 LF 精炼轴承钢选择渣洗

Components (by mass)/% 化学成分(质量分数)/%				
CaO	SiO_2	Al_2O_3	R	ΣFeO
50~60	13~18	15~20	≥3.5	≤1.0

Through the reasonable control of LF refining process, the quality of bearing steel is effectively improved. According to the national standard of bearing steel GB 18254—2002, the oxygen content is specified clearly: the oxygen content in die cast steel is required to be no more than 15ppm and the oxygen content in continuous cast steel is required to be no more than 12ppm. This goal can be achieved by LF refining process. The oxygen content control level can reach 7~10ppm, and the minimum oxygen content can reach 6.8ppm to meet the production requirements of bearing steel.

通过 LF 精炼工艺的合理控制,有效地提高轴承钢的质量。按照轴承钢国家标准 GB 18254—2002对氧含量做了明确规定:要求模铸钢中的氧含量不大于 15ppm,连铸钢中氧含量不大于 12ppm。利用 LF 精炼工艺可以实现此目标,氧含量的控制水平可以达到 7~10ppm,最低氧含量可达到 6.8ppm,能够满足轴承钢的生产要求。

2.4.6 Other New Technologies of LF Refining Process

2.4.6 其他 LF 精炼工艺新技术

With the development of iron and steel industry, steelmaking and secondary refining technologies for steel products with special needs, high specifications and high requirements, as well as high added value, are continuously developed, including wide and thick plates, high alloy oil well pipes, high difficulty wire rod products, etc. The development of these products provides technical support for the development of LF Secondary Refining, such as deep deoxidation and inclusion control technology, including inclusion denaturation technology to meet the needs of steel, inclusion control technology with size no more than 20μm, ladle refining slag modification technology, refining technology to reduce the total amount of inclusions, etc.; deep desulfurization refining technology, making [S] in steel no more than 0.0010%; precise control technology of molten steel composition and temperature in refining process; refining process control the development and application of the model; the production technology of clean steel; the control technology of nitrogen in refining process; the research on the development and improvement of service life of the refractories used in the secondary refining and clean steel production; the high-precision, rapid analysis and detection means to meet the requirements of clean steel, etc., develop and optimize the LF refining technology around the needs of special products to form a set of refining equipment and technology. At present, the main features of these new technologies include:

随着钢铁产业发展,炼钢及炉外精炼工艺技术需要开发满足特殊需要的、高规格、高要求同时也是高附加值的钢铁产品,包括宽厚板、高合金油井管材、高难度盘条产品等,这些产品的开发为 LF 炉外精炼技术的开发提供技术支撑。例如深脱氧和夹杂物控制技术,包括满足钢种需要的夹杂物变性技术、尺寸不大于 20μm 的夹杂物控制技术,钢包精炼渣改质技术,减少夹杂物总量的精炼工艺技术等;深脱硫精炼工艺技术,使得钢中 [S] ≤ 0.0010%;精炼过程中钢水成分和温度的精确控制技术;精炼过程控制模型的开发与应用;洁净钢生产技术;精炼过程氮的控制技术;耐材使用寿命的研究;高精度、快速分析和检测手段等。开发并优化围绕特殊产品需要的 LF 精炼工艺技术,形成套精炼设备和工艺技术。目前这些新技术主要的特点包括:

(1) High efficiency and high speed: The development of converter and continuous casting process is aimed at high speed, and refining is often the 'Bottleneck' in steelmaking process. Due to the limitation of temperature rise and desulfuration speed, the production rhythm of LF furnace is difficult to meet the requirements of high-efficiency converter or high-speed continuous casting.

(1) 高效化和高速化:转炉和连铸工艺的发展均以高速化为目标,精炼往往为炼钢生产流程中的"瓶颈"。LF 炉工艺受升温和脱硫速度的限制,生产节奏已很难适应高效转炉或高速连铸的要求。

(2) Online configuration of rapid analysis facilities: Accurately predict the end point and temperature of molten steel refining, select the best refining process and use computer to control the operation of oxygen blowing, stirring, feeding, alloy adjustment, molten steel heating and temperature control in the refining process. Therefore, the uncertainty of raw materials and operation in the smelting process makes online measurement more and more important.

(2) 在线配置快速分析设施:准确预报钢水精炼的终点与温度,选择最佳的精炼工艺,并利用计算机控制精炼过程中的吹氧、搅拌、加料、合金调整与钢水加热和温度控制等操作。因此,冶炼过程的原料及操作不确定性使得在线检测变得越来越重要。

(3) Intelligent control: For example, the intelligent oxygen determination technology is used to realize the oxygen and aluminum determination in LF refining furnace, the rapid oxygen and aluminum determination in argon blowing station, the control of oxygen activity and acid melting aluminum in LF, and the determination of slag (FeO) in LF.

(3) 智能化控制:例如运用智能化定氧技术可以实现 LF 精炼炉定氧定铝、吹氩站快速定氧定铝、LF 内氧活度和酸熔铝的控制及 LF 内炉渣 (FeO) 的测定等。

Here, two examples of more mature new technologies applied in some production enterprises are given as follows:

这里列举两项在一些生产企业中运用较为成熟的新技术:

(1) The nitrogen increasing control technology of LF furnace: The molten metal of LF furnace is heated by high temperature of electric arc. Arc is a kind of high-temperature and high-speed gas jet. Its impact on the solution tank is essentially similar to that of the oxygen stream in the converter, which creates a pit at the impact point. When the electrode is heated, the exposed

liquid steel will appear in the pit, and the temperature of the exposed liquid steel is higher than that of the other parts, which is more than 2400K. At this temperature, the surface activity of O and S hindering the nitrogen absorption of the liquid steel disappears. At this time, it is easy to absorb nitrogen as long as the liquid steel exposed. Under the action of high temperature of arc, the nitrogen in the surrounding air is basically dissociated into a single atom state, which also creates conditions for the liquid steel to absorb nitrogen. To solve this problem, the nitrogen increasing control technology of LF furnace is used. The brief introduction is as follows:

（1）LF 炉增氮控制技术：LF 炉熔池金属靠电弧的高温加热。电弧是一种高温高速的气体射流，对溶池的冲击作用和转炉中氧气流股的冲击作用在本质上是相似的，即在冲击点处造成一个凹坑。电极加热时，凹坑处会出现裸露的钢液面，这部分裸露的钢液较其他部位的钢液温度高，大于 2400K。该温度下，O、S 对阻碍钢液吸氮的表面活性作用消失，此时只要钢液裸露就容易吸氮。在电弧的高温作用下，周围空气中的氮气基本上全部被离解为单原子状态，这也给钢液吸氮创造了条件。运用 LF 炉增氮控制技术能够解决以下问题：

1) The foam residue is made as early as possible: The steel surface (especially at the impact pit) is prevented from being exposed. The thickness of foam slag layer should be higher than the depth of impact pit. Some domestic steel mills have implemented foam slag process in LF. 15 furnace foam slag tests were carried out on high strength ship plate steel. Two batches of foaming agent were added during electrode heating in each furnace, with an interval of 7min. The content of nitrogen, oxygen and acid soluble aluminum in the steel sample is analyzed when the molten steel is moved into and out of LF furnace. The amount of nitrogen added to molten steel is 1~13ppm, with an average of 5.27ppm. Three to four batches of foaming agent should be added according to the length of heating and refining time for steel grades with strict control of nitrogen content.

1) 尽早造好泡沫渣：防止钢液面（尤其是冲击凹坑处钢液面）裸露。泡沫渣渣层厚度应高于冲击凹坑深度。国内部分钢厂已在 LF 炉实施泡沫渣工艺，在高强船板钢种上做了 15 炉泡沫渣试验。每炉试验在电极加热期间加入 2 批发泡剂，间隔 7min。在钢水进入 LF 精炼和出 LF 精炼时取钢样分析钢中氮、氧、酸溶铝含量。钢水增氮量为 1~13ppm，平均增氮 5.27ppm。对严格控制钢中氮含量的钢种应根据加热和精炼时间的长短加入 3~4 批发泡剂。

2) Control oxygen in liquid steel: There is no aluminum in the refining process. In the production of low nitrogen steel, aluminum wire and titanium should be fed at the later stage of LF refining. For steel grades requiring VD vacuum refining, aluminum and titanium should be added 5min before vacuum breaking by alloy adding device under vacuum.

2) 控制钢液中的氧：精炼中间不加铝。生产低氮含量钢时应在 LF 精炼后期喂铝线加铝和钛。需要 VD 真空精炼的钢种应通过真空下合金加入装置，在破真空前 5min 加入铝、钛。

（2）Erosion control technology of LF refractory: Several factors causing lining erosion of ladle furnace are as follows:

（2）LF 耐材侵蚀控制技术：造成钢包炉炉衬蚀损的因素包括：

1) The corrosion of slag line refractories will be accelerated significantly during refining. Due to the mutual inductance between the three-phase electrodes, the arc in the electrode system repels each other. In the ladle furnace, the radiation energy and the convection energy are first aligned in three directions, forming an angle of 120°. The superheated slag is pushed to impact the ladle wall, forming the erosion of the slag on the ladle wall. The deep is 100~150mm. High basicity slag can reduce the influence of the radiation energy and stabilize the arc.

1）精炼时会显著加速渣线耐火材料的损蚀。由于三相电极之间的互感，电极系统内的电弧相互排斥。在钢包炉中，辐射能和对流能先对准三个方向，呈120°，过热渣被推动冲击钢包壁，形成炉渣对包壁的冲刷，深的100~150mm。高碱度渣能减少辐射能的影响，并且能够稳定电弧。

2) Affected by arc flicker, local heating and cooling produce stress on the refractory, which leads to peeling and slag penetration into the joint of lining brick, and the joint cracks under stress.

2）受电弧闪烁影响，局部加热和冷却在耐材上产生应力，导致剥落和炉渣渗入衬砖接合处，在应力作用下接合处开裂。

3) Gas agitation will aggravate the corrosion of refractories.

3）气体搅拌会加剧耐火材料的蚀损。

The related technologies to protect the lining of ladle furnace by using LF refining to control the erosion of refractory include:

利用LF精炼实现耐材侵蚀控制技术保护钢包炉炉衬的相关技术包括：

(1) Using functional bulk refractories, the hot gunning technology of ladle slag line is carried out.

（1）采用功能性散装耐火材料，进行钢包炉渣线热喷补技术。

(2) Slag foaming technology.

（2）炉渣的泡沫化技术。

(3) The refining slag technology with low oxygen level, high sulfide capacity and weak compatibility with MgO refractories is formed.

（3）形成氧位低、硫化物容量高，并与MgO耐火材料相容性弱的精炼渣技术。

(4) Slag tapping technology to reduce the entrainment of oxidizing slag into ladle furnace, such as EBT electric arc furnace.

（4）留渣出钢技术，减少氧化性渣夹带进入钢包炉，如EBT电弧炉。

2.4.7 Limitations of LF Refining Process

2.4.7 LF炉精炼工艺的局限性

LF is the highest refining equipment at present. LF was originally used to solve the problem of long production cycle of EAF, and it has the best matching effect with EAF. At present, LF refining is often used in converter production process. However, due to the influence of production cycle mismatch, the burden of LF furnace is too large and the time of reducing slag and desulfuration is not enough. Therefore, the limitations of LF refining process need to be made up by reducing the amount of slag discharged from converter, strengthening deoxidation and LF refining slag optimization. The solutions adopted by some iron and steel enterprises include:

LF 是目前使用最高的精炼设备。LF 原是为了解决电弧炉生产周期长，与电弧炉搭配效果最佳。现转炉生产流程也常采用 LF 精炼，但受到生产周期不匹配的影响，LF 炉负担太大，造还原渣和脱硫时间不够，因此还需要通过减少转炉出钢下渣量、加强脱氧和 LF 精炼渣优化，弥补 LF 炉精炼工艺的局限性。一些钢铁企业采用的解决方案包括：

(1) The primary furnace should increase the tapping temperature properly, and strengthen the ladle baking, turnover and heat preservation.

(1) 初炼炉应适当提高出钢温度，加强钢包烘烤、周转和保温。

(2) Adding slag modifier, premelted slag and bottom blowing during tapping move the refining process forward; Adding refining slag along with the steel flow during tapping to realize LF slag making and deoxidizing moving forward, and adopting low melting point premelted slag to realize rapid slag making.

(2) 出钢时加炉渣改质剂、预熔渣和底吹使精炼过程前移；出钢时炉渣改质或同时随钢流加入精炼渣，实现 LF 炉造渣和脱氧前移，采用低熔点的预熔渣实现快速造渣。

(3) Optimize the power supply, make good submerged arc slag, and improve the heating speed; Improve the ladle temperature, good insulation and high power supply.

(3) 优化供电，造好埋弧渣，提高升温速度；提高钢包温度，保温良好，以及大功率供电。

(4) Improve the automation level, shorten the sampling and analysis time, and improve the control hit rate of composition and temperature.

(4) 提高自动化水平，缩短取样分析时间，提高成分与温度的控制命中率。

(5) One primary furnace is equipped with two LFS, or two primary furnaces are equipped with three LF.

(5) 1 座初炼炉配 2 座 LF，或 2 座初炼炉配 3 座 LF。

(6) For steelmaking plants with more than two casters, the method of parallel smelting of low-grade steel and high-quality variety steel can be adopted, so that only the process route of argon blowing station can be followed, and only high-quality variety steel can be used for LF.

(6) 对于有 2 台以上铸机的炼钢厂，可以采用低档钢种与优质品种钢平行冶炼的方式生产，这样可以只走吹氩站的工艺路线，LF 只走优质品种钢。

In order to reduce the hydrogen content of molten steel, the humidity of slag must be strictly controlled to keep dry. In particular, lime is easy to absorb moisture, and the use of passivated lime is better than that of ordinary lime. In addition, the slag quantity should be strictly controlled, which is generally 2% ~ 8% (by mass) of the metal quantity. Before refining, the oxidized slag of converter or electric furnace must be removed as much as possible, and then new slag can be produced.

为了降低钢液的含氢量，必须严格控制渣料的湿度，使之保持干燥。特别是石灰，易吸潮，使用钝化石灰比一般石灰好。另外，要严格控制渣量，一般渣量为金属量的2%~8%（质量分数）。在精炼之前必须尽可能地将转炉或电炉氧化性炉渣扒掉，然后造新渣。

In addition, the hydrogen content of molten steel refined by LF furnace can not be controlled, the main reasons include：

另外，单纯经过LF炉精炼的钢水氢含量却无法实现控制，主要原因包括：

(1) Atmospheric water is the main factor of hydrogen absorption. Especially in the process of tapping in contact with the atmosphere, hydrogen is absorbed by slag and diffused to the molten steel sheet. Hydrogen will also be absorbed during the addition of slag and alloy.

(1) 大气水分是吸氢的主要因素。特别是在出钢过程与大气接触，氢通过渣吸入并扩散至钢液。在渣料和合金加入过程中也会吸氢。

(2) The addition of lime, dolomite and other solvents will increase the hydrogen content in the steel. Lime is easy to absorb moisture in the atmosphere, which often causes high hydrogen content.

(2) 添加石灰、白云石等溶剂会增加钢中的氢含量。石灰容易吸收大气中的湿气，这是经常造成氢含量高的主要原因。

(3) Alkaline slag tends to keep water.

(3) 碱性渣保持水分的倾向大。

(4) When adding ferroalloy, hydrogen will increase, and the alloy will absorb some water, resulting in the increase of hydrogen content.

(4) 加铁合金时氢会增加，合金会吸收一些水分造成氢含量增加。

(5) Oxygen or sulfur content can reduce hydrogen absorption, while low oxygen and low sulfur steel can absorb hydrogen more easily.

(5) 氧或硫含量高时能减少吸氢，低氧低硫钢更容易吸氢。

(6) High temperature increases the solubility of hydrogen in molten steel, and calcium treatment increases hydrogen.

(6) 高温会增加氢在钢液中的溶解度，钙处理后会增氢。

Therefore, it is necessary to use vacuum degassing device, inert gas stirring and other comprehensive means to complete the refining task of molten steel.

因此，需要通过应用真空脱气装置、惰性气体搅拌等综合手段完成钢液的精炼任务。

Task 2.5　LF Refining Equipment
任务2.5　LF炉精炼设备

Ladle furnace is a ladle refining equipment developed in Japan in the late 1970s. LF ladle furnace is composed of ladle body, bottom blowing argon permeable brick, sliding nozzle, alloy funnel and electrode heating system attached to the furnace cover. LF ladle furnace is widely used because of its simple equipment, low investment cost, flexible operation and good refining effect. Because of its simple structure, various metallurgical functions and flexibility in use, the LF equipment has significant refining effect and high economic benefits, and has become an important equipment in the iron and steel production process. According to incomplete charging, China has more than 200 LF furnaces of different sizes.

LF 炉是 20 世纪 70 年代末期日本发展起来的钢包精炼型炉外精炼设备。LF 钢包炉由钢包炉体、包底吹氩透气砖、滑动水口、炉盖上附有合金漏斗和电极加热系统等组成。因 LF 钢包炉设备简单，投资费用低，操作灵活和精炼效果好而受到普遍重视，并得到广泛的应用。由于 LF 设备结构简单，具有多种冶金功能和使用中的灵活性，精炼效果显著，具有较高的经济效益，成为钢铁生产流程中的重要设备。据不完全统计，我国已拥有 LF 炉 200 台以上。

2.5.1 Equipment Composition of LF Furnace

2.5.1 LF 炉的设备组成

LF furnace is an external refining method characterized by electric arc heating and slag making refining. The main equipment of LF furnace (shown in Figure 2-5) includes furnace body (ladle), electric arc heating system, alloy and slag feeding system, bottom blowing argon stirring system, wire feeding system, furnace cover and cooling water system (some without cooling system), dust removal system, temperature measurement and sampling system, ladle car control system, etc. According to the way of power supply, it can be divided into alternating current ladle furnace and direct current ladle furnace. Direct current ladle furnace includes single electrode direct current ladle furnace, double electrode direct current ladle furnace and three electrode direct current arc electroslag ladle furnace. At present, most of the domestic furnaces use alternating current ladle furnace.

LF 炉是以电弧加热和造渣精炼为主要技术特征的炉外精炼方法。LF 炉主要设备（见图 2-5）包括炉体（钢包）、电弧加热系统、合金与渣料加料系统、底吹氩搅拌系统、喂线系统、炉盖及冷却水系统（有的没有冷却系统）、除尘系统、测温取样系统、钢包车控制系统等。按照供电方式分为交流钢包炉和直流钢包炉，直流钢包炉包括单电极直流钢包炉、双电极直流钢包炉和三电极直流电弧电渣钢包炉。目前国内多数使用交流钢包炉。

The basic layout is as follows:

基本布局形式为：

(1) Two station rotary tables mode: a station refining, and a station soft argon blowing (waiting).

(1) 双工位回转台方式：一个工位精炼，一个工位软吹氩（等待）。

(2) A processing position+A hoisting position mode.

(2) 一个处理位+一个吊包位方式。

(3) A processing position+Two hoisting and wrapping position mode.

(3) 一个处理位+两个吊包位方式。

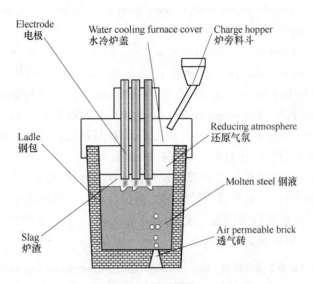

Figure 2-5 Equipment of LF refining furnace
图 2-5 LF 精炼炉设备

2.5.1.1 LF Furnace Body

2.5.1.1 LF 炉炉体

The ladle is the main equipment of LF process system. The upper opening of the ladle has a water-cooled flange, which is sealed with the furnace cover through a sealing rubber ring to prevent air intrusion. When the ladle is used for vacuum treatment, it is also required that its shell is welded with steel plate according to the airtight welding conditions. At the bottom of the ladle, there is a sliding nozzle for steel pouring and a permeable brick for argon blowing at the position $r/2 \sim r/3$ (r is the ladle inner radius) from the furnace wall. The argon flow rate in refining process is determined by different working positions and ladle capacity. The flow rate of argon can reach $3 \sim 4L/(min \cdot t)$ to stir the molten steel. The lining is magnesia carbon brick, magnesia chrome brick, magnesia alumina spinel, high alumina brick, or zirconia chrome brick. According to the process requirements of refined steel, the comprehensive bricklaying method is adopted.

钢包是 LF 工艺系统的主体设备，这种钢包的上口有水冷法兰盘，通过密封橡皮圈与炉盖密封，防止空气侵入。当钢包用于真空处理时，还要求其外壳用钢板按气密焊接条件焊成。钢包底部有浇钢用的滑动水口及距炉壁，在 $r/2 \sim r/3$ (r 为钢包内半径) 处设有吹氩用的透气砖。精炼过程中氩气流量根据不同工位和钢包容量等决定。氩气流量可达 $3 \sim 4L/(min \cdot t)$，以达到搅拌钢液的目的。包衬为镁碳砖、镁铬砖、镁铝尖晶石、高铝砖或锆铬砖。根据精炼钢种的工艺要求，采用综合砌砖法。

The ratio H/D of molten pool depth H to molten pool diameter D in LF ladle is a factor that must be considered in ladle design. The H/D value of ladle furnace affects the stirring effect of molten steel, the contact area of slag, the heat load of slag line, the life and heat loss of ladle lining, etc. Generally, the bath depth H of refining furnace is relatively large, and H/D is 0.9 ~ 1.3. The distance from the liquid level of steel to the opening of ladle is called the free space of ladle furnace. For the ladle without vacuum treatment, the height of the free space is smaller. For the ladle without vacuum treatment, the free space is generally 500 ~ 600mm. For the ladle with vacuum treatment, the free space is generally 800 ~ 1200mm or even 1500mm or higher. The H/D values of LF furnaces with different capacities are listed in Table 2-6.

LF 炉钢包内熔池深度 H 与熔池直径 D 之比 H/D 是钢包设计时必须要考虑的因素。钢包炉的 H/D 数值影响钢液搅拌效果、钢渣接触面积、包壁渣线带的热负荷、包衬寿命和热损失等。一般精炼炉的熔池深度 H 都比较大，H/D 为 0.9 ~ 1.3。从钢液面至钢包口的距离称为钢包炉的自由空间。对非真空处理用的钢包，自由空间的高度小一些，自由空间一般为 500 ~ 600mm；真空处理的钢包一般为 800 ~ 1200mm，有的甚至达 1500mm 或更高。不同容量 LF 炉的 H/D 值见表 2-6。

Table 2-6 H/D value of LF furnace with different capacity

表 2-6 不同容量的 LF 炉 H/D 值

Capacity/t LF 炉容量/t	20	30	60	150	50
Loading/t 实际装入量/t	13/23	18/33	60	100/150	45/50
Out diameter /mm 外径/mm	2200	2400	2600	3900	2924
Inter D/mm 内径 D/mm	1676	1948	2070	3164	2430
High/mm 总高/mm	2300	2500	3150	4330	3040
Inter high/mm 内高/mm	1995	2195	2740	4000	2770
Bath depth H/mm 熔池深 H/mm	1260	1402	2340	2754	1348
H/D	0.75	0.72	1.13	0.87	0.49

2.5.1.2 Arc Heating System

2.5.1.2 电弧加热系统

The electric arc heating system equipment used in LF furnace is basically the same as that of electric arc furnace, which is composed of furnace transformer, short net, electrode lifting mechanism, conductive cross arm and graphite electrode. The electric arc produced between three graphite electrodes and molten steel is used as a heat source to heat the molten steel, because the electrode is inserted into the slag through the furnace lid hole, so it is called submerged arc heating. The heating method has less heat dissipation, and reduces the heat radiation and erosion of arc light on the furnace lining. It can also stabilize the current. Compared with electric furnace, the method of submerged arc heating can adopt lower secondary voltage. The heating rate of molten steel can reach 4℃/min.

LF炉所使用的电弧加热系统设备也与电弧炉基本相同，由炉用变压器、短网、电极升降机构、导电横臂和石墨电极所组成。三根石墨电极与钢液间产生的电弧作为热源加热钢液，由于电极通过炉盖孔插入泡沫渣中，故称埋弧加热。此种加热法散热少，能减少电弧光对炉衬热辐射和侵蚀，可稳定电流。采用埋弧加热方法，与电炉相比，更能使用较低的二次电压。钢液升温速度可达4℃/min。

The secondary side of the transformer used in LF furnace is usually divided into several levels of voltage, but it is not necessary to adjust the load on load. Because there are many switching modes when there is no load, the equipment is simple and cheap, and the reliability is good. During LF refining, the liquid level of steel is relatively stable, and the current fluctuation is small. When there is no electric furnace to melt the furnace charge, the allowable current density is relatively large due to the short-circuit impact current caused by the collapse of the charge. Power supply parameters of LF furnaces with different furnace capacities are shown in Table 2-7.

LF炉所用的变压器，其副边通常也分为数级电压，但没有必要进行有载调荷。因为无载时切换方式很多，设备简单便宜，可靠性好。LF炉精炼时钢液面比较平稳，电流波动较小。而一般电弧炉熔化废钢时，由于塌料所引起的短路冲击电流，电流密度需要很大。不同炉容的LF炉供电参数见表2-7。

Table 2-7　Power supply parameters of LF furnace with different furnace capacities
表2-7　不同炉容的LF炉供电参数

Capacity t/month 炉容 t/month	20 2000	30 2500	60 4500	150 4000	50 1500
Transformer/kV·A 变压器/kV·A	3000	5000	6500	6000	7500
Secondary voltage/V 二次电压/V	220/95	235/85	225/75	275/110	250/102
Secondary current/A 二次电流/A	10600	14400	23860	17000	17320
Diameter of electrode/mm 电极直径/mm	254	254	356	356	305
Diameter of electrode circle/mm 电极圆直径/mm	650	600	810	940	900
Per ton steel power/kV·A·t^{-1} 吨钢功率/kV·A·t^{-1}	150	167	108	40	150

LF furnace uses submerged arc heating with low voltage and high current to refine molten steel. The electrode regulating system should adopt the automatic regulating system with good reaction and high sensitivity. The electrode lifting speed of LF furnace is generally 2~3m/min. In order to avoid the thermal radiation of the arc to the ladle lining, the three electrodes are arranged in a compact way.

LF 炉采用低电压、大电流埋弧加热法进行钢液精炼。电极调节系统要采用反应良好、灵敏度高的自动调节系统。LF 炉的电极升降速度一般为 2~3m/min。为避免电弧对钢包衬的热辐射,三根电极采用紧凑式布置。

In the early stage of heating, low power supply is used for slag making; after slag melting, submerged arc refining is realized and high power supply is used. After continuous heating for 10~15min, the heating is stopped for 2~5min, so as to make the temperature of the molten steel uniform up and down, and not to cause excessive local temperature on the slag surface. The heating time is about 30min. The heating rate is controlled at 2~4℃/min. The energy distribution and heat transfer between the electrode and the molten steel during the heating process are shown in Figure 2-6.

在加热初期,采用低功率供电造渣;渣料熔化后,实现埋弧精炼,采用高功率供电。连续加热 10~15min 后,应停止加热 2~5min,以便使钢液温度上下均匀,不会造成渣面局部温度过高。加热时间约 30min,升温速度控制在 2~4℃/min。加热过程的能量分布及电极与钢液传热如图 2-6 所示。

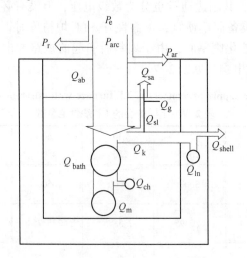

Figure 2-6　Energy distribution during LF furnace heating
图 2-6　LF 炉加热过程的能量分布

P_e—Active power output by transformer; P_r—Electric energy lost by the line (short network); P_{arc}—Arc power; P_{ar}—Arc power lost; Q_{ab}—Heat of electric arc entering slag steel molten pool; Q_{bath}—Heat energy retained in the molten pool; Q_{ch}—Used for the temperature rise of slag and alloy melting; Q_m—Temperature rise heat of molten steel and slag; Q_k—Heat loss through furnace lining; Q_{ln}—Heat storage of furnace lining; Q_{sa}—Heat loss from slag surface; Q_g—Heat taken away by furnace gas; Q_{sl}—Heat emitted from slag surface; Q_{shell}—Heat lost by heat exchange between the cladding and the surrounding atmosphere

P_e—变压器输出的有功功率;P_r—线路(短网)损失的电能;P_{arc}—电弧功率;P_{ar}—损失的电弧功率;Q_{ab}—进入渣钢熔池中的电弧热量;Q_{bath}—滞留在熔池中的热能;Q_{ch}—用于渣料、合金熔化升温热;Q_m—钢水、炉渣的升温热;Q_k—通过炉衬损失的热量;Q_{ln}—炉衬的蓄热;Q_{sa}—由渣面损失的热量;Q_g—炉气带走的热量;Q_{sl}—由渣面散发出的热量;Q_{shell}—由包壳与周围大气的热交换而损失的热量

2.5.1.3 Water Cooling Furnace Cover of LF

2.5.1.3 LF 炉水冷炉盖

The furnace cover is the key part of LF ladle refining furnace design, because in many cases the metallurgical effect of ladle refining furnace depends on the atmosphere control in the furnace to a great extent. In order to prevent air from entering the furnace from the gap between the furnace cover and the ladle and the opening of the furnace cover, some necessary protective measures have been taken, including cleaning and maintenance of the ladle opening which must be taken during the use. For example, the cover of 120t ladle refining furnace is full water-cooled tube type (as shown in Figure 2-7). The whole cover is welded with seamless steel pipe and special elbow, forming an efficient water-cooled forced circulation with uniform flow and no dead point.

炉盖是 LF 钢包精炼炉设计的关键部分,因为很多情况下钢包精炼炉的冶金效果在很大程度上取决于炉子内的气氛控制。为了避免空气从炉盖和钢包之间的间隙及炉盖开孔处进入炉内,采取了一些必要的保护措施,包括使用过程中必须采取包口清理和维护工作。例如,120t 钢包精炼炉的炉盖为全水冷管式炉盖(见图 2-7),整个炉盖用无缝钢管和特制弯头组焊而成,形成均流无死点的高效水冷强制循环。

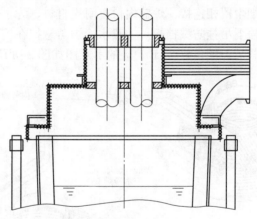

Figure 2-7　120t LF furnace cover system
图 2-7　120t LF 炉炉盖系统

The LF furnace cover is used for sealing the ladle mouth, maintaining a strong reducing atmosphere in the furnace, preventing the ladle from radiating heat and improving the heating efficiency. LF furnace cover is of water-cooled structure. The furnace cover is lined with refractory. In order to prevent the adhesion between the furnace cover and the ladle caused by the splashing of molten steel, some of them hang a splash proof baffle under the furnace cover. The whole water-cooled furnace cover is hung on the door hanger with adjustable chain hook at four points, and there is a lifting mechanism on the hanger. The position of the furnace cover can be adjusted as required. For LF furnace with vacuum degassing system, in addition to the heating cover mentioned above, LF furnace with vacuum system (LFV furnace) also has a vacuum furnace cover, which is connected with the vacuum system for degassing molten steel. Alloy feeding port, slag feeding device and temperature measuring (or sampling device) are arranged on both furnace covers of LF furnace.

LF 炉盖用于钢包口密封，以及保持炉内强还原性气氛，从而防止钢包散热，提高加热效率。LF 炉盖为水冷结构，炉盖内层衬有耐火材料。为了防止钢液喷溅而引起的炉盖与钢包的黏连，有的炉盖下面还吊挂一个防溅挡板。整个水冷炉盖在四个点上用可调节的链钩悬挂在门形吊架上，吊架上有升降机构，可根据需要调整炉盖的位置。有真空脱气系统的 LF 炉，除上述加热盖以外，还有一个真空炉盖，与真空系统相连，用来进行钢液脱气。在 LF 炉的两种炉盖上都设有合金加料口、渣料加料装置和测温（或取样）装置。

LF furnace cover is generally water-cooled tube type to ensure the strong reducing atmosphere in the furnace, prevent the heat dissipation of the ladle and improve the heating efficiency. The water-cooled furnace cover is connected with the smoke exhaust dust cover, and the furnace cover is lined with refractory materials. In order to prevent the adhesion between the furnace cover and the can body caused by the splashing of liquid steel, the anti splashing baffle is hung under the furnace cover. Generally, the water-cooled furnace cover is hung on the door-shaped hanger, and there is a lifting mechanism on the hanger. The position of the furnace cover can be adjusted as required. The schematic diagram of LF furnace cover is shown in Figure 2-8.

LF 炉盖一般是管式水冷的，以保证炉内的强还原气氛，防止钢包散热，从而提高加热效率。水冷炉盖和排烟尘罩相连接，炉盖内衬有耐火材料。为了防止钢液喷溅而引起炉盖与桶体的黏连，在炉盖下吊挂防溅挡板。一般水冷炉盖悬挂在门形吊架上，吊架上有升降机构，可根据需要调整炉盖的位置。LF 炉盖的示意图如图 2-8 所示。

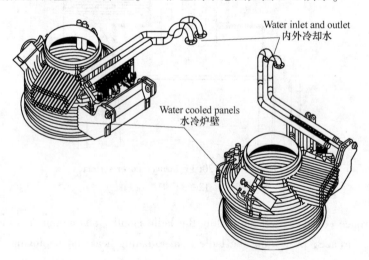

Figure 2-8　LF furnace cover
图 2-8　LF 炉盖

The furnace cover structure includes:

炉盖结构包括：

(1) Three electrode holes.

(1) 三个电极孔。

(2) Dedusting hole, connected with dedusting system.

(2) 除尘孔，与除尘系统连接。

(3) Automatic feeding hole.

(3) 自动加料孔。

(4) An observation hole is opened at the side of the furnace cover or at the edge of the top surface, which is convenient for manual feeding and temperature measurement sampling, and it can also be used to observe the smelting situation in the furnace.

(4) 炉盖侧面或在顶面边部开有观察孔，便于人工加料及测温取样，同时还可通过它观察炉内冶炼情况。

(5) Feeding hole, feeding aluminum wire for deoxidization or feeding calcium silicate wire for inclusion denaturation.

(5) 喂线孔，喂铝线脱氧或喂硅钙线进行夹杂物变性处理。

(6) If there is an automatic temperature measurement sampling device, an automatic temperature measurement sampling hole is also opened on the furnace cover.

(6) 如果有自动测温取样装置，炉盖上还应开有自动测温取样孔。

Each return water of the water-cooled furnace cover is respectively equipped with a thermal resistance to monitor the return water temperature, so as to reflect the working condition of each cooling part and facilitate adjustment and treatment.

水冷炉盖各路回水分别设有监测回水温度的热电阻，以反映各冷却部位的工作状况，便于调整和处理。

2.5.1.4 Feeding System

2.5.1.4 加料系统

The LF furnace is generally equipped with alloy and slag hopper on the heating cover. The furnace materials weighed by the electronic scale enter the ladle furnace through the chute and feeding port. LF furnace with vacuum system is generally equipped with alloy and slag charging device on the vacuum cover. Its structure is basically the same as that on the heating cover, except that a vacuum sealing valve should be added at each joint, as shown in Figure 2-9.

LF 炉一般在加热包盖上设合金及渣料料斗。通过电子秤称量过的炉料，经溜槽、加料口进入钢包炉内。有真空系统的 LF 炉，一般在真空盖上设合金及渣料的加料装置。其结构与加热包盖上的基本上相同，只是在各接头处均需加上真空密封阀，如图 2-9 所示。

2.5.1.5 Slag Raking Device

2.5.1.5 扒渣装置

One of the refining functions of LF furnace is to produce reductive white slag. Therefore, before LF refining, the oxidized slag must be removed. Therefore, LF furnace must have the function of slag removal. There are two ways to remove slag:

LF 炉精炼功能之一是造还原性白渣精炼。为此，在 LF 炉精炼之前，必须将氧化性炉渣去掉。因此，LF 炉必须具备除渣功能。除渣的方式有两种：

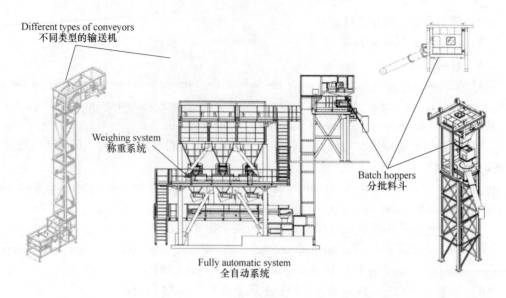

Figure 2-9 LF furnace charging system
图 2-9 LF 炉加料系统

(1) When LF furnace adopts multi station operation, tilting and slag raking device can be set on ladle car. When the ladle car drives to the slag raking station, the slag raking operation can be carried out.

(1) 当 LF 炉采用多工位操作时，可在放钢包的钢包车上设置倾动、扒渣装置。当钢包车开到扒渣工位时，即可进行扒渣操作。

(2) If LF furnace adopts fixed position and moving furnace cover, ladle tilting device shall be set on LF furnace base. Before refining, slag shall be removed first, new slag added, and then refining heated.

(2) 如果 LF 炉采用固定位置，在炉盖移动形式时，则需把钢包倾动装置设在 LF 炉底座上，在精炼前先扒渣，再加新渣料，然后再加热精炼。

2.5.1.6 Powder Spraying and Wire Feeding Device

2.5.1.6 喷粉及喂线装置

In LF refining, powder spraying equipment is often used to desulfurate, purify and microalloy molten steel. The powder spraying equipment includes a ladle cover, a spray gun for powder spraying, a powder spraying tank and a powder bin. During powder spraying, the powder is weighed and mixed automatically, and then sent to the powder spraying tank through the screw feeder. High purity argon is used as carrier gas during powder spraying, and the flow rate is 200~400L/min. Generally, the treatment time is 5~10min. The wire feeding device is shown in Figure 2-10.

LF 炉精炼时常采用喷粉设备对钢液进行脱硫、净化和微合金化等操作。喷粉设备包括钢包盖、一支喷粉用的喷枪和喷粉罐和粉料料仓。喷粉时对粉料先自动称重及混合，然后通过螺旋给料器送至喷粉罐。喷粉时采用高纯氩气作载流气，流量为 200~400L/min，通常处理时间为 5~10min。喂线装置如图 2-10 所示。

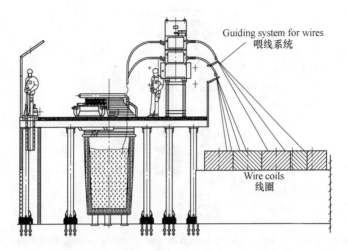

Figure 2-10　LF furnace wire feeding device
图 2-10　LF 炉喂线装置

2.5.1.7　Temperature Measurement and Sampling System

2.5.1.7　测温取样系统

LF refining furnace is generally equipped with automatic temperature measurement and sampling facilities, as shown in Figure 2-11. Automatic temperature measurement can achieve fixed-point temperature measurement, and the measured temperature value is more representative, which can avoid the fluctuation of human factors on the temperature measurement of molten steel. At the same time, the use of white dynamic temperature measurement and sampling system also reduces the labor load of workers. But at present, even if LF has automatic temperature measurement equipment, most of the production plants still use manual temperature measurement due to the habit and possible failure of the equipment during field operation, which may lead to the result that the temperature value of molten steel measured by different operators has some deviation.

LF 精炼炉一般都配有自动测温取样设施，如图 2-11 所示。自动测温可实现定点测温，所测温度值更具有代表性，可避免人为因素对钢液温度测量产生的波动。同时使用自动测温取样系统，减轻了工人的劳动负荷。目前 LF 即使有自动测温设备，但由于现场操作时的习惯及设备可能出现故障等原因，大部分生产厂仍采用人工测温，这样有可能出现不同操作者测出的钢液温度值有所偏差的结果。

2.5.1.8　Ladle System

2.5.1.8　钢包车系统

The ladle car is used to transport the ladle. The movement of the ladle car is driven by the

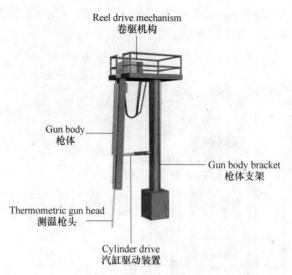

Figure 2-11　Temperature measurement and sampling device of LF furnace
图 2-11　LF 炉测温取样装置

motor through the reducer and the coupling to transmit the torque by the wheels, two of which are the driving wheels and the other two are the driven wheels, which can achieve a wide range of speed adjustment and obtain a large torque in a large range. The high-speed shaft is equipped with a brake and a travel switch. The main power line, control line and argon gas pipe are sent to the ladle through 'Flexible Hose' and sliding wire device. The control system of ladle car mainly includes operation, stop, limit control and weighing. Ladle and ladle car system are shown in Figure 2-12.

钢包车用来运送钢包，钢包车的运动是由电机经减速器、联轴器带动车轮传递扭矩。两个为主动轮，另外两个为从动轮，可以实现大范围的速度调节，在很大范围内获得大转矩。高速轴处设有制动器，并设有行程开关。总的电源线、控制线及氩气管是通过"软线软管"及滑线装置送到钢包车上的。钢包车控制系统主要包括运行、停止、限位控制及称量。钢包及钢包车系统如图 2-12 所示。

2.5.1.9　Dedusting System

2.5.1.9　除尘系统

The smoke and dust are collected to the main dedusting system through LF dedusting pipe and treated by the plant in a centralized way.

烟尘通过 LF 除尘管汇集到主除尘系统，由厂统一集中处理。

2.5.2　LF Furnace Electrode

2.5.2　LF 炉电极

The LF furnace electrode is the core equipment to generate heat. Through the contact between the electrode and the molten steel, the current collides to generate heat, and the heat generated is

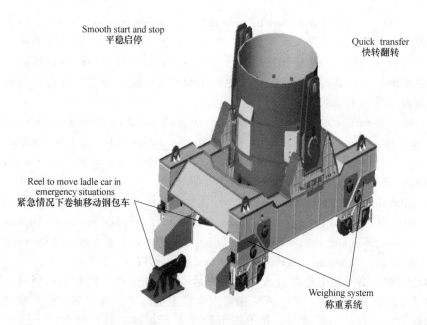

Figure 2-12　Ladle car device of LF furnace
图 2-12　LF 炉钢包车装置

transmitted to the molten steel by the electron. In addition, the radiation effect of the current during the discharge process will also generate heat, as shown in Figure 2-13.

　　LF 炉电极是产生热量的核心设备。通过电极与钢液接触，电流冲撞产生热量，由电子将产生的热量传至钢液，另外放电过程中电流的辐射作用也会产生热量，如图 2-13 所示。

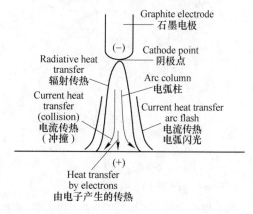

Figure 2-13　Heat distribution during LF electrode discharge
图 2-13　LF 电极放电过程热量分布

　　The consumption of electrode is inevitable in the process of using. Electrode consumption can be divided into normal consumption and accidental consumption. Normal consumption includes end consumption caused by arc behavior, erosion of steel and slag liquid, and side consumption caused by oxidation. Accidental consumption includes electromagnetic stress, high temperature thermal shock, improper operation, and fracture of top end of joint hole, fracture of pin center and fracture of low end of joint hole caused by poor physical performance of electrode itself.

电极使用过程中的消耗是无可避免的。电极消耗分为正常消耗和意外消耗两种,正常消耗包括电弧行为及钢、渣液侵蚀冲刷而造成的端部消耗,氧化导致的侧面消耗;意外事故消耗包括电磁应力,高温热冲击,以及电极本身物理性能差导致的接孔顶端断裂,销子中心断裂以及接孔低端断裂。

The most common accidental consumption is electrode breakage. Electrode fracture is a sudden accident in the operation of LF furnace, which can be divided into high-level fracture and low-level fracture. High level fracture usually occurs at the highest joint or joint seat of the electrode column, which is the joint part of the electrode holder. Low level fracture is the fracture of the lower end of the electrode (especially at the joint), which is often thin due to local oxidation and under the action of external force. In this case, it is usually knocked out manually first, in order to prevent the electrode end from accidentally falling into the molten pool and increasing carbon. The influence of the arc on the electrode in use is shown in Figure 2-14.

最常见的意外消耗是电极断裂。电极断裂是 LF 炉操作中突然发生的事故,可分为高位断裂和低位断裂两种情况。高位断裂通常发生在电极柱的最高接头或接头座处,也就是电极把持器的接续部位。低位断裂是电极下端(尤其在接头处)经常由于局部氧化而变细,在外力作用下发生断裂。在这种情况下,为了避免电极端头无意中掉进熔池增碳,通常先被人工打掉。电极在使用过程的电弧影响如图 2-14 所示。

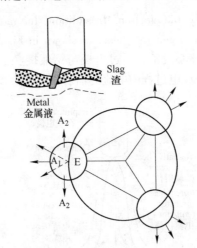

Figure 2-14 Arc behavior
图 2-14 电弧行为

A_1—Main area of end consumption caused by arc behavior; A_2—Secondary area of end consumption caused by arc behavior; E—Steel and slag erosion area

A_1—电弧行为引起端部消耗主要区域;A_2—电弧行为引起端部消耗的次要区域;E—钢、渣侵蚀区域

2.5.3 LF Refining Equipment Improvement

2.5.3 LF 精炼的设备改进

In order to meet the needs of smelting more steel grades, and make full use of the secondary refining means, aiming at the cost, quality and other problems in the use process of LF refining furnace, the production enterprise continuously improves the LF refining equipment in combination with the actual situation. There are three kinds of brief introduction:

为了满足冶炼中更多钢种的需要,充分利用炉外精炼手段针对 LF 精炼炉使用过程中成本、质量等问题,对 LF 精炼的设备进行不断改进。精炼手段包括:

(1) NK-AP method: NK-AP method, developed in 1981 in Japan NKK Fukuyama Iron Making Institute, uses plug-in gas gun instead of permeable brick to carry out gas mixing and spraying of refined powder, as shown in Figure 2-15.

(1) NK-AP 法:1981 年,NK-AP 法在日本 NKK 福山制铁所开发成功,其使用插入式气体喷枪代替透气砖,进行气体搅拌和精炼粉剂的喷吹,如图 2-15 所示。

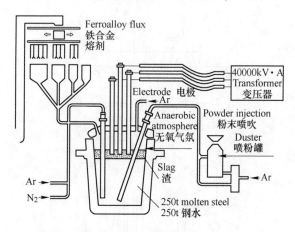

Figure 2-15　NK-AP equipment diagram
图 2-15　NK-AP 设备图

(2) PLF method: In this way, the heating in LF furnace is changed into plasma gun with graphite electrode. Because there is no dissolution of carbon stripped off by carbon electrode, it is very effective for the smelting of very low carbon steel. PLF method was first used in Nippon Steel in 1993, as shown in Figure 2-16.

(2) PLF 法:此方法把 LF 炉中加热用的石墨电极换成等离子体枪。由于没有碳电极剥落而导致碳的溶解,因此对于极低碳钢的冶炼是非常有效的。1993 年,PLF 法首次在日本新日铁用于生产,其结构如图 2-16 所示。

(3) Multifunctional LF method: In order to realize the LF refining and dehydrogenation, the ladle is placed in the vacuum tank, and the multi-functional LF method of argon stirring, arc heating, powder injection and alloy element addition is set up. As shown in Figure 2-17, the combination of ladle degassing and LF refining is called LVF method, which is introduced in detail after VD vacuum dehydrogenation refining.

(3) 多功能 LF 法:为了实现 LF 精炼脱氢,生产企业把钢包放在真空槽中,设置出吹氩搅拌、电弧加热、喷吹精炼粉剂、添加合金元素等设备的多功能 LF 法,其设备如图 2-17 所示。把钢包脱气和 LF 精炼组合在一起,称为 LVF 法,此方法在 VD 真空脱氢精炼后予以详细介绍。

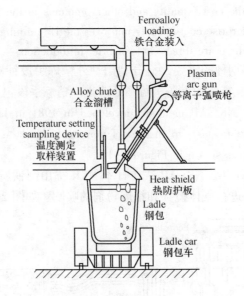

Figure 2-16 PLF equipment
图 2-16 PLF 设备

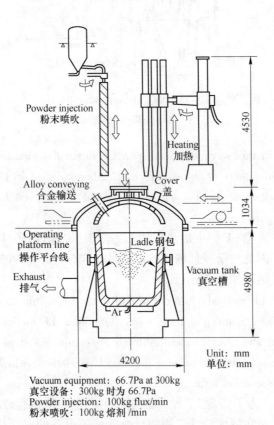

Vacuum equipment: 66.7Pa at 300kg
真空设备: 300kg 时为 66.7Pa
Powder injection: 100kg flux/min
粉末喷吹: 100kg 熔剂/min

Figure 2-17 LFV multifunctional refining equipment
图 2-17 LFV 多功能精炼设备

Task 2.6 ASEA-SKF Ladle Refining
任务 2.6 ASEA-SKF 钢包精炼

ASEA-SKF is a ladle refining furnace developed by Swedish Ball Bearing Company (SKF) and Swedish General Electric Company (ASEA) in 1965. It is the combination of electromagnetic stirring vacuum degassing equipment and electric arc furnace. It has perfected three basic functions of modern refining equipment: heating, vacuum and stirring for the first time. ASEA-SKF ladle refining furnace uses electromagnetic stirring, vacuum degassing, electric arc heating and slag washing refining and other means to complete tasks including electric arc heating, vacuum degassing, vacuum oxygen blowing decarburization, desulfurization, and further adjustment of composition and temperature, and oxygen removal and inclusion removal under electromagnetic induction stirring of molten steel.

ASEA-SKF 是瑞典滚珠轴承公司（SKF）与瑞典通用电气公司（ASEA）于 1965 年研制成功的钢包精炼炉。它是电磁搅拌真空脱气设备与电弧炉的结合，第一次完善了现代炉外精炼设备的三个基本功能，即加热、真空和搅拌。ASEA-SKF 钢包精炼炉运用电磁搅拌、真空脱气、电弧加热和渣洗精炼等手段，可以完成的任务包括电弧加热、真空脱气、真空吹氧脱碳、脱硫以及在电磁感应搅拌钢液下进一步调整成分和温度、脱氧和去夹杂等。

2.6.1 ASEA-SKF Ladle Refining Furnace Process Operation

2.6.1 ASEA-SKF 钢包精炼炉工艺操作

When the molten steel is tapped, the ladle furnace is hoisted into the mixer for electromagnetic induction mixing. At this time, the primary slag is removed, and the slag material is added to make new slag. When the molten steel temperature is appropriate, the vacuum cover is put on the back cover for vacuum degassing treatment. After vacuum degassing, alloy is added into the chute funnel to adjust the composition of molten steel. Finally, the molten steel is heated to a proper temperature, and then the ladle is lifted out for pouring. The whole refining process is completed in about 1.5~3h. The process setting is shown in Figure 2-18.

当钢水出钢后，将钢包炉吊入搅拌器内进行电磁感应搅拌，此时除掉初炼渣。加渣料造新渣，待钢水温度合适后盖上真空盖，进行真空脱气处理。真空脱气后，通过斜槽漏斗加入合金调整钢水成分，最后将钢水加热到合适温度，再将钢包吊出进行浇注。整个精炼过程约在 1.5~3h 内完成。工艺设置如图 2-18 所示。

ASEA-SKF ladle refining furnace has two main process arrangements: fixed furnace body and movable furnace body, as shown in Figure 2-19. The furnace body fixed type refers to the fixed position of electromagnetic stirring plate and ladle, and the vacuum furnace cover and heating furnace cover can rotate and rise and fall. The furnace body moving type means that the ladle and the

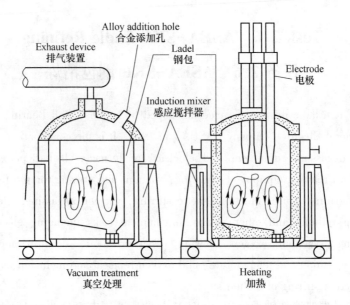

Figure 2-18　Process setting of ASEA-SKF ladle refining furnace
图 2-18　ASEA-SKF 钢包精炼炉工艺设置

mixing plate are placed on a rail car, while the vacuum cover and the heating cover are fixed at their respective positions, which can only be lifted and lowered, not rotated.

ASEA-SKF 钢包精炼炉的工艺布置主要有两种，其分别为炉身固定式和炉身移动式，如图 2-19 所示。炉身固定式是指电磁搅拌片和钢包位置固定，真空炉盖和加热炉盖可以旋转、升降；炉身移动式是指钢包和搅拌片放在一个轨道车上，而真空盖和加热盖在各自的位置固定，只能升降，不能旋转。

2.6.2　ASEA-SKF Ladle Refining Furnace Equipment Structure

2.6.2　ASEA-SKF 钢包精炼炉设备结构

ASEA-SKF ladle refining furnace equipment structure consists of: ladle containing molten steel, arc heating system, vacuum sealed furnace cover and vacuum pumping system, slag and alloy feeding system, oxygen blowing system, argon blowing stirring system, control system. It also includes the auxiliary equipment of slag removal equipment, vacuum temperature measurement, sampling equipment etc.

ASEA-SKF 钢包精炼炉设备结构组成包括：盛装钢水的钢包，电弧加热系统，真空密封炉盖和抽真空系统，渣料及合金加料系统，吹氧系统，吹氩搅拌系统和控制系统。同时还包括除渣设备、真空测温取样设备等辅助设备。

Electromagnetic Stirring
电磁搅拌

The main characteristics of ASEA-SKF ladle refining furnace is electromagnetic stirring. ASEA-SKF ladle refining furnace adopts electromagnetic induction stirring, and the equipment is

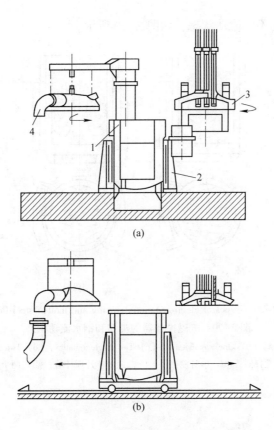

Figure 2-19 ASEA-SKF furnace layout
图 2-19 ASEA-SKF 炉布置方式
(a) Furnace cover rotary type; (b) Ladle mobile type
(a) 炉盖旋转式；(b) 钢包移动式
1—Ladle; 2—Agitator; 3—Electrode cover; 4—Vacuum cover
1—钢包；2—搅拌器；3—电极盖；4—真空盖

composed of transformer, low frequency converter and induction mixer. Generally, the transformer adopts oil immersed natural cooling three-phase transformer. The secondary current of the transformer is sent to the frequency converter through the water-cooled cable. The frequency converter adopts the silicon controlled low frequency converter, which can adjust the frequency automatically or manually. The stirring frequency is generally controlled at 0.5~1.5Hz, and the movement speed of molten steel is generally controlled at about 1m/s. There are mainly two types of agitators: cylinder agitator and chip agitator. The different arrangement of the induction mixer can control the different flow states of the liquid steel, as shown in Figure 2-20.

电磁搅拌是 ASEA-SKF 钢包精炼炉最主要的特点。ASEA-SKF 钢包精炼炉采用电磁感应搅拌，设备由变压器、低频变频器和感应搅拌器组成。变压器一般采用油浸式自然冷却三相变压器，经过水冷电缆将变压器二次电流送给变频器。变频器一般采用可控硅式低频变频器，通过自动或手动方式调整频率。搅拌频率一般控制在 0.5~1.5Hz，钢液运动速度一般控制在 1m/s 左右。搅拌器主要有圆筒式搅拌器和片式搅拌器两种。感应搅拌器的不同布置可以控制钢液的不同流动状态，如图 2-20 所示。

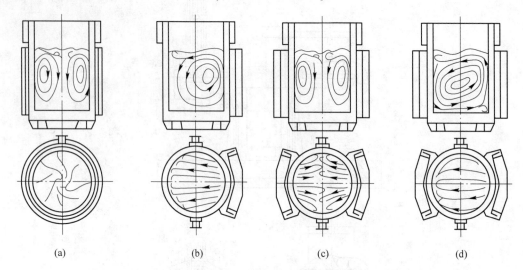

Figure 2-20　Types of electromagnetic stirring and molten steel flow state

图 2-20　电磁搅拌器的类型和钢水流动状态

（a）Drum agitator；（b）One way agitator；（c）Two single agitators；（d）Two agitators in series

（a）圆筒式搅拌器；（b）单向搅拌器；（c）两个单片搅拌器；（d）两个搅拌器串联

Ladle

钢包

ASEA-SKF ladle refining furnace ladle shell is made of non-magnetic steel plate. The ladle has a certain free space. The shape requirement of ladle is that the stirring force decreases rapidly with the increase of the ratio of ladle diameter to stirring coil height. The thickness of the lining refractory is generally 230mm. The distance between agitator and ladle shall be as small as possible.

ASEA-SKF 钢包精炼炉钢包外壳由非磁性钢板构成。钢包留有一定的自由空间。钢包的形状要求是：随着钢包直径与搅拌线圈高度比值的增大，搅拌力迅速降低。包衬耐火材料的厚度一般为 230mm。搅拌器与钢包之间的距离应尽可能小。

Heating System

加热系统

The heating system includes transformer, electrode furnace cover, electrode arm, electrode and its lifting system. The heating power is lower than that of LF. The heating furnace cover is made according to the common electric arc furnace cover.

加热系统包括变压器、电极炉盖、电极臂、电极以及其升降系统。与 LF 相比，加热功率要低一些。加热炉盖按照普通电弧炉盖制作。

Vacuum System

真空系统

The vacuum system is a vacuum chamber composed of a sealed furnace cover and a ladle. After the ladle is heated, remove the furnace cover, move the ladle together with the agitator under the vacuum cover, and cover the vacuum cover for vacuum treatment. The structure of the vacuum system is shown in Figure 2-21.

真空系统是由一个密封炉盖与钢包一起构成的一个真空室。当钢包加热结束之后，移开加热炉盖，将钢包连同搅拌器一起移至真空盖下，然后盖上真空盖进行真空处理。其真空系统结构如图 2-21 所示。

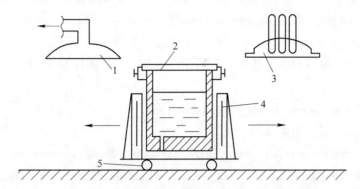

Figure 2-21　ASEA-SKF ladle refining vacuum and arc heating system
图 2-21　ASEA-SKF 钢包精炼真空及电弧加热系统
1—Vacuum furnace cover; 2—Ladle; 3—Heating furnace cover; 4—Electromagnetic mixing system; 5—Ladle car
1—真空炉盖；2—钢包；3—加热炉盖；4—电磁搅拌系统；5—钢包车

2.6.3　Refining Effect of ASEA-SKF Ladle Refining Furnace

2.6.3　ASEA-SKF 钢包精炼炉精炼效果

As a ladle refining technology, ASEA-SKF ladle refining furnace combines LF electric arc heating and VD vacuum degassing to integrate heating, vacuum and agitation. The refining effects which can be achieved by using ASEA-SKF ladle refining furnace include:

作为钢包精炼技术，ASEA-SKF 钢包精炼炉是综合 LF 电弧加热和 VD 真空脱气两种工艺的结合，将加热、真空、搅拌这三种精炼手段融为一体。运用 ASEA-SKF 钢包精炼炉能够实现的精炼效果包括：

(1) Improve production efficiency: The operation rhythm of steelmaking production and the refining speed is improved, and the original turnover time is reduced in the vacuum system and heating system.

(1) 提高生产效率：提高了炼钢生产的运转节奏和精炼速度，减少原本在真空系统和加热系统周转时间。

(2) Improve the quality of steel: The vacuum system can be used to realize the degassing function of ASEA-SKF ladle refining furnace, and the refining quality of steel can be improved from the following four aspects:

(2) 提高钢质量：ASEA-SKF 钢包精炼炉可以利用真空系统实现钢水脱气功能，提高钢液精炼质量可分为以下四个方面：

1) Gas content: It can reduce hydrogen in molten steel to less than 0.0002% and the content of oxygen to 40%~60%.

1) 气体含量：可使钢水中的氢降至 0.0002% 以下，氧含量降低 40%~60%。

2) Inclusions in steel: Basically eliminate the low-power inclusions and improve the high-power inclusions.

2) 钢中夹杂物：基本消除低倍夹杂，高倍夹杂明显改善。

3) Mechanical properties: the fatigue strength is increased. The energy is impacted by 10%~20%, and elongation and reduction of area by 10% and 20%.

3) 力学性能：疲劳强度与冲击功提高 10%~20%，延伸率和断面收缩率提高 10% 和 20%。

4) Cutting performance: It has great improvement.

4) 切削加工性能：很大改进。

(3) The production is stable, which makes the ladle refining process safer and more stable. ASEA-SKF ladle refining furnace does not need a long and specific turnover ladle, and only needs to be carried out in the same station, so that the whole refining process is more safe and stable.

(3) 生产平稳，使钢包精炼过程更加安全稳定。ASEA-SKF 钢包精炼炉不需要较长具体的周转钢包，只需要在相同工位上进行，使整个精炼流程更加安全稳定。

2.6.4 Evaluation of ASEA-SKF Ladle Refining Furnace

2.6.4　ASEA-SKF 钢包精炼炉的使用评价

The ASEA-SKF ladle refining furnace of 30t of Fushun Special Steel, Liaoning Province, China, has been put into production since the end of 1990s, with stable production, reliable refining process mixing and low operating cost. Short arc is allowed for induction mixing, with less heat radiation to the ladle lining. The power factor is large, and the thermal efficiency of the arc is high. The advantages of its using process include:

自 20 世纪 90 年代末投入生产以来，中国辽宁抚顺特钢 30t ASEA-SKF 钢包精炼炉生产平稳，精炼过程搅拌可靠，操作费用低；感应搅拌允许使用较短电弧，对包衬热辐射少；功率因数较大，电弧的热效率较高。ASEA-SKF 钢包精炼炉的优点包括：

(1) It can be used for arc heating and electromagnetic induction stirring of molten steel. The degassing time of molten steel can be unlimited. Inclusions are easy to float into slag. and the operation is flexible. It can be used for deoxidization, desulfurization, composition and temperature adjustment. The quality of molten steel can be greatly improved.

(1) 可对钢水进行电弧加热和电磁感应搅拌，钢水的脱气时间可不受限制，夹杂物易于上浮到渣中，操作灵活，可进行脱氧、脱硫、调整成分和温度，钢水质量可大幅度提高。

(2) It can improve the productivity of primary smelting furnace, which only melts.

(2) 可提高初炼炉的生产率，初炼炉一般只起熔化作用。

(3) The variety can be expanded. Because a large number of ferroalloys are added in the re-

fining process, many kinds of steel can be produced from carbon steel to alloy steel.

（3）可扩大品种。由于精炼过程中加入了大量铁合金，因而可生产由碳素钢到合金钢等很多品种。

(4) It can reduce production cost. As the smelting time of primary smelting furnace greatly shortened, the power consumption is reduced, the alloy yield is increased, and the surface quality of ingot is improved, so the amount of ingot grinding is less. In addition, the hydrogen diffusion annealing time of large section billet can be reduced, so the energy consumption can also be reduced.

（4）可降低生产成本。由于初炼炉的熔炼时间大大缩短，降低了电耗，提高了合金收得率，同时还改善了钢锭的表面质量。

However, the use process also revealed some shortcomings, mainly including：

但是在使用过程中也暴露出一些缺点及不足，其主要包括：

(1) The equipment is complex, the low-frequency power supply is used for mixing, the cost is high, and the refining time is long. The stirring of molten steel near the induction coil is stronger than other parts, the erosion of furnace lining is more serious, and the consumption of refractory materials is high (shown in Figure 2-22).

（1）设备较复杂，搅拌用低频电源，造价高，精炼时间也较长；在感应线圈附近钢水搅动较其他部位强烈，对炉衬的侵蚀也较为严重，耐火材料消耗高，如图2-22所示。

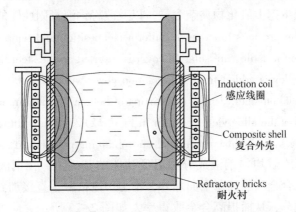

Figure 2-22　Schematic diagram of induction coil of ASEA-SKF ladle refining furnace
图2-22　ASEA-SKF钢包精炼炉感应线圈部位示意图

(2) The equipment investment is large, the steel flow is stable in the mixing process, which can not make the molten steel and slag mix well, and the desulfurization effect is poor. Therefore, in order to solve this problem, the ASEA-SKF ladle refining furnace has been equipped with a low blow gas stirring system to improve the stirring effect in the refining process.

（2）设备投资大，搅拌过程中钢流稳定，不能使钢水与炉渣进行很好混合，脱硫效果较差。为解决此问题，目前抚钢ASEA-SKF钢包精炼炉已配加低吹气体搅拌系统，以提高精炼过程的搅拌效果。

Task 2.7　CAS Refining
任务 2.7　CAS 精炼

In the late 1980s, the simple out of furnace treatment technology developed by the Bafan Technology Research Institute of Nippon Steel Corporation of Japan was gradually popularized and applied. CAS refining has the functions of homogenizing the composition and temperature of molten steel, improving the yield of alloy, purifying molten steel, and removing gas and inclusions. It is a simple ladle refining technology without vacuum facilities, which can blow argon while adding alloy from the surface of molten steel in a closed vessel excluding ladle slag.

CAS——密封吹氩微调合金成分。日本新日铁公司八幡技术研究所1975年开发的简易炉外处理工艺，20世纪80年代末期逐渐推广应用。CAS精炼均匀钢液成分和温度，提高合金收得率，净化钢液，去除气体和夹杂物的功能。CAS精炼是一种无真空设备的简单钢包精炼技术，在钢包渣以外的密闭容器中，需在钢水表面加入合金的同时吹氩。

Under atmospheric pressure, the argon treatment of molten steel can achieve the goal of uniform temperature and composition, accelerating the melting of alloy and deoxidizer, reducing the content of oxide inclusions in steel and improving the solidification performance of molten steel.

在大气压下，通过钢包吹氩进行钢液氩气处理，可实现均匀温度和成分，加速合金料和脱氧剂的熔化，减少钢中氧化物夹杂含量，以及改善钢液凝固性能的目的。

On the basis of the previous function of uniform composition and temperature of molten steel by bottom blowing argon in ladle, an isolation cover is inserted on the steel liquid surface, the slag is removed, and ferroalloy is added from the isolation cover without contacting with the slag, directly added into the molten steel. Under the stirring of bottom blowing argon, the fine-tuning alloy composition improves the alloy yield (shown in Figure 2-23). In addition, this method is also benefical to eliminating large inclusions in steel and obtaining pure liquid steel at low cost.

在以往的钢包底吹氩均匀钢液成分和温度功能的基础上，在钢液面上插入一个隔离罩，撇开炉渣，让铁合金从隔离罩上面加入，不与炉渣接触，直接加到钢液内，并在底吹氩搅拌下微调合金成分，从而提高合金收得率，如图2-23所示。此外，这种方法还有利于消除钢中的大型夹杂物，能够以较低的成本获得纯净的钢液。

2.7.1　CAS Process Flow

2.7.1　CAS 工艺流程

During the treatment of CAS, argon is first blown from the bottom of the ladle to form a slag free area on the surface of the molten steel. Under the action of rising Gas-Liquid two-phase flow stream, the slag is pushed around to form a slag free area on the surface of the molten steel, and then the alloy is added. As the cover is sealed, there is no slag cover on the steel surface. In addition, argon blowing makes the cover full of argon, forming a relatively inert atmosphere. Com-

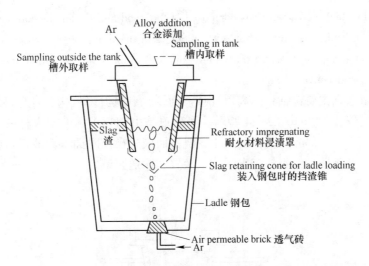

Figure 2-23 CAS refining process
图 2-23 CAS 精炼工艺

pared with the conventional argon blowing and stirring treatment, the total oxygen content in the steel can be reduced from 100ppm to 40ppm after 7~8min of treatment. After treatment, the non-metallic inclusions in the steel are obviously reduced and the quality of the steel is greatly improved.

进行 CAS 处理时，首先要从钢包底吹氩气，在钢水表面形成一个无渣的区域，在上升的气液两相流股作用下，将熔渣向周围推开，钢液表面形成一个无渣区域，然后加入合金。由于罩内呈密封状态，钢液面无炉渣覆盖，加上吹氩使罩内充满氩气，形成相对的惰性气氛。与常规吹氩搅拌处理钢液比较，处理 7~8min 后，钢中总氧量由 100ppm 降到 40ppm 以下，处理后钢中非金属夹杂物明显减少，从而大大改善钢的质量。

The ladle is lifted to the treatment station. And after alignment, the argon is forced to blow for 1min. After the slag layer on the surface of the molten steel is blown open, the hood immediately lowered, the temperature measured, the sample taken at the same time, the calculated alloy weighed, the argon blown continuously, and the ferroalloy can be added later for stirring. At the end of argon blowing, lift the isolation cover, and then measure the temperature for sampling.

具体操作流程为：钢包吊运到处理站，对位以后，强吹氩 1min。吹开钢液表面渣层后，立即降罩，同时测温取样，按计算好的合金称量，不断吹氩，稍后即可加入铁合金进行搅拌。吹氩结束将隔离罩提升，再测温取样。

2.7.2 CAS Refining Equipment

2.7.2 CAS 精炼设备

The argon blowing system at the bottom of the ladle mainly includes: the isolation hood (or immersion hood) and its lifting mechanism, alloy charging system, temperature measurement sampling, and slag cutting device (shown in Figure 2-24). The isolation hood is a refractory ring hood supported by steel plate, which is composed of the hood and the immersion part. The immersion part is cast with high alumina refractory cement, and the liquid steel is inserted during operation. The alloy charging system includes alloy bin electromagnetic vibration feeder, electronic weighing device, and conveyor belt.

钢包底吹氩系统主要包括：隔离罩（或浸渍罩）和其升降机构，合金料加入系统，以及测温取样和割渣装置，如图 2-24 所示。隔离罩是由钢板支承的耐火材料环罩，由罩体和浸入部分组成，浸入部分用高铝耐火水泥浇铸而成，操作时插入钢液。合金料加入系统包括合金料仓、电磁振动给料器、电子称量装置和输送胶带。

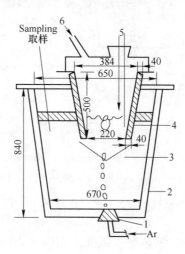

Figure 2-24　CAS refining equipment diagram
图 2-24　CAS 精炼设备示意图
1—Permeable brick；2—Ladle；3—Liquid steel；4—Isolation cover；5—Heating and sampling hole；6—Exhaust hole
1—透气砖；2—钢包；3—钢液；4—隔离罩；5—加热及取样孔；6—排气孔

2.7.3　CAS Refining Function and Effect

2.7.3　CAS 精炼功能及效果

The main functions of CAS refining include: uniform composition and temperature of molten steel, adjustment of composition and temperature of molten steel (scrap cooling), improvement of alloy yield (especially aluminum), and purification of molten steel and removal of inclusions. The refining effects can be achieved as follows:

CAS 精炼主要功能包括：均匀钢水成分和温度，调整钢水成分和温度（废钢降温），提高合金收得率（尤其是铝），以及净化钢水，去除夹杂物。能够实现精炼效果为：

(1) The total oxygen content in steel decreased from 0.0100% to less than 0.0040% after conventional argon blowing treatment；

(1) 钢中总氧量由常规吹氩处理的 0.0100% 下降到 0.0040% 以下；

(2) The large inclusions above 40μm can be reduced by 80%, and the inclusions of 20~40μm can be reduced by 1/3~1/2.

(2) 40μm 以上的大型夹杂物可减少 80%，20~40μm 的夹杂物可减少 1/3~1/2。

(3) The yield of alloy can be increased by 20%~50% by increasing the yield of deoxidizing elements. The yield of aluminum in Baosteel increased from 40% of conventional deoxidization (10%~30% in general plants) to 80%~90%, and that of titanium from 50%~80% of conventional deoxidization is to about 100%.

(3) 提高脱氧元素的收得率，可提高合金收得率 20%~50%。宝钢铝的收得率由常规脱氧的 40%（一般厂 10%~30%）提高到 80%~90%，钛由常规脱氧的 50%~80% 提高到约 100%。

2.7.4 CAS-OB Refining

2.7.4 CAS-OB 精炼

CAS-OB refining is to add oxygen gun on the basis of CAS equipment. In order to quickly compensate the temperature drop in CAS refining process, an oxygen lance is added in the isolation hood to blow oxygen to the molten steel, and aluminum or ferrosilicon is added to the molten steel at the same time. The molten steel is heated directly by the heat generated by the reaction of the added aluminum (or ferrosilicon) with oxygen, which is called CAS-OB (OB: Oxygen Blowing). Its purpose is to raise the temperature of converter liquid steel rapidly, compensate the temperature drop of CAS process, provide accurate target temperature for the molten steel in tundish, and make converter and continuous casting coordinate. CAS-OB method is the most representative chemical heating method, which was first introduced by Nippon Steel.

CAS-OB 精炼是在 CAS 设备的基础上增设氧枪。为了快速补偿 CAS 精炼处理过程中的温降，在隔离罩内增设了一支吹氧枪对钢液进行吹氧，同时向钢液内加入铝或硅铁，利用加入的铝（或硅铁）与氧反应，放出的热量直接加热钢液，称为 CAS-OB 法。其目的是对转炉钢液进行快速升温，补偿 CAS 法工序的温降，为中间包内的钢液提供准确的目标温度，使转炉和连铸协调配合。CAS-OB 法最早为新日铁推出，是最具代表性的化学加热法。

2.7.4.1 CAS-OB Refining Principle

2.7.4.1 CAS-OB 精炼原理

CAS-OB refining technology is developed on the basis of argon blowing at the bottom of ladle, and the function of OB is added to CAS. It separates the exposed molten steel from the air and most of the ladle covering slag by immersing it in a closed immersion hood, and gradually forms an inert atmosphere in the hood by blowing argon at the bottom, which lays a good foundation for the external refining treatment of molten steel, such as composition regulation, molten steel purification, oxygen blowing combustion temperature rise, etc. Under atmospheric pressure, argon is blown in through the permeable brick at the bottom of the ladle. The immersion tube is inserted above the argon blowing port on the molten steel surface, and the slag is prevented. The composition of molten steel can be adjusted by adding various alloys into the immersion tube. And an oxygen gun is inserted on the immersion tube to heat the molten steel. CAS-OB can adjust and even the temperature and composition of molten steel, reduce the content of oxide inclusions in steel and improve the solidification performance of molten steel.

CAS-OB 精炼技术是在钢包底吹氩的基础上发展起来的，是对 CAS 加装了 OB 功能。它通过浸入密闭的浸罩到钢水中将裸露的钢水与空气和大部分钢包覆盖渣隔开，并通过底吹氩，在罩内逐步形成惰性气氛，为钢水成分调节、钢水净化、吹氧燃烧升温等炉外精炼的处理打下良好的基础。在大气压下，将氩气通过钢包底部透气砖吹入，将浸渍管插入钢液表面吹氩口的上方，并挡掉炉渣。可通过浸渍管加入各种合金调整钢水成分，并且在浸渍管上插入氧枪，加热钢水。CAS-OB 可以调整和均匀钢水的温度和成分，减少钢中氧化物夹杂的含量，以及改善钢水的凝固性能。

Process Principle of CAS-OB Method
CAS-OB 法的工艺原理

The basic principle of CAS-OB method include: add top oxygen gun in the isolation hood to blow oxygen, and use the heat generated by the reaction of aluminum (or ferrosilicon) added in the hood and oxygen to directly heat the molten steel. Oxygen is blown through the OB oxygen gun, and aluminum, silicon, carbon and other combustible materials are put into the cover to generate chemical heat. The heat of combustion is transmitted to the molten steel by radiation, conduction and convection, and it is transferred to the deep part of the molten steel by argon stirring. The theoretical oxygen demand of burning 1.0kg aluminum is $0.62m^3$, and that of burning 1.0kg silicon is $0.8m^3$. The oxygen demand of CAS-OB method is $0.74m^3$ and $1.05m^3$ respectively. The specific heat capacity of liquid steel is $0.88kJ/(kg \cdot ℃)$, and the heating values of 1.0kg aluminum and 1.0kg silicon per ton steel are 35℃ and 33℃ respectively. Generally, the temperature rise range is controlled within 50℃, and the heat absorption rate can be more than 80%. In order to obtain better loss of iron, manganese and carbon, it is necessary to keep more theoretical combustion materials in equilibrium with oxygen supply.

CAS-OB 法的基本原理为：在隔离罩内增设顶氧枪吹氧，利用罩内加入的铝（或硅铁）与氧反应所放出的热量直接对钢水加热。通过 OB 氧枪吹氧，并向罩内投入铝、硅、碳等可燃物质，从而发生氧化反应产生化学热。燃烧热通过辐射、传导、对流传给钢液，借助氩气搅拌将热传向钢液深部。燃烧 1.0kg 铝的理论需氧量为 $0.62m^3$，燃烧 1.0kg 硅需要氧 $0.8m^3$；而使用 CAS-OB 法时，需氧量分别为 $0.74m^3$ 和 $1.05m^3$。钢液比热容取 $0.88kJ/(kg \cdot ℃)$，燃烧 1.0kg 铝和 1.0kg 硅的吨钢升温值分别为 35℃ 和 33℃。一般把升温幅度控制在 50℃ 以内，热吸收率可达 80% 以上。为获得较好的铁、锰、碳元素烧损，需要保持与供氧相平衡的理论燃烧物数量。

CAS-OB Refining Process Heating
CAS-OB 精炼过程加热

In CAS-OB process, aluminum (or ferrosilicon) is continuously added into the molten steel through the isolation hood while oxygen blowing, and the molten steel is heated by the chemical heat of aluminum (or silicon) oxidation. During oxygen blowing, a small amount of manganese, carbon and iron elements are oxidized, and the heating rate is 5 ~ 10℃/min. For example, Baosteel's maximum heating speed can reach 13℃/min. When the tapping temperature (or the pouring temperature) is low, it can be smelted without returning to the furnace after oxygen blowing treatment.

CAS-OB 法在吹氧的同时, 不断通过隔离罩向钢液内加入铝（或硅铁）, 利用铝（或硅）氧化的化学热对钢液加热。吹氧时有少量锰、碳、铁元素被氧化, 升温速度为 5 ~ 10℃/min。如宝钢最大升温速度可达 13℃/min。当出钢温度（或浇注温度）低时, 经吹氧处理, 可以不回炉再冶炼。

The purpose of CAS-OB method to heat and regulate the temperature of molten steel is to better coordinate steelmaking and continuous casting. The ladle treatment station acts as a buffer station to provide accurate target temperature for the molten steel in the tundish. For example, the ANS-OB of Angang Steel is controlled at the target temperature of ±3℃. As the heating in the process is realized, it is helpful to reduce the tapping temperature of converter and improve the furnace life and liquid steel quality. Moreover, it can coordinate the production of converter and continuous casting, promote the continuous casting of multiple furnaces and improve the productivity. It is easy to operate low cost and high efficiency to heat the molten steel directly by the reaction of aluminum (or ferrosilicon) with oxygen.

CAS-OB 法对钢液加热调温的目的是为了使炼钢与连铸能更好地协调配合。钢包处理站起到缓冲作用, 为连铸中间包内的钢液提供准确的目标温度。例如鞍钢的 ANS-OB, 温度控制在目标温度±3℃。由于实现了过程中加热, 有助于降低转炉的出钢温度, 提高炉龄和钢液质量。并且可以协调转炉与连铸的生产, 促进多炉连浇, 提高生产率。利用加入的铝（或硅铁）与氧反应所放出的热量直接加热钢液, 操作方便, 且成本低, 效率高。

CAS-OB adds the function of argon blowing and stabilize in the original CAS station. The oxygen gun is installed in the center of the isolation cover, and the top blowing self consuming oxygen gun is used. Aluminum and other alloys are directly put into the steel water surface from the feeding port, and the exothermic agent is mainly aluminum. When the temperature is raised, the continuous aluminum supply method is used, and the heating rate is 5~6℃/min.

CAS-OB 在原 CAS 站增设吹氩提稳功能。增设的氧枪安装在隔离罩的中心, 并采用顶吹自消耗型氧枪, 铝及其他合金由加料口直接投到钢水面, 放热剂主要是铝。提温时采用连续供铝方式, 升温速度 5~6℃/min。

Argon Stirring in CAS-OB Refining Process
CAS-OB 精炼过程吹氩搅拌

In CAS refining treatment, the immersion hood is inserted into the molten steel to isolate the air from the steel level after slag removal. As the bottom blown argon enters into the immersion hood through the molten steel, the air in the hood is gradually discharged through the upper flue gas pipeline to form an argon atmosphere in the immersion hood. For CAS refining, the 'Vacuum' reaction zone of argon is formed in the bubble generated by bottom blowing argon and in the immersion hood above the liquid level of steel. Because CAS refining has the function of preventing air from entering and no secondary oxidation, a large amount of argon is blown into the bottom of ladle, so that the refining function of argon bubble metallurgy can be fully exerted, as shown in Figure 2-25.

在 CAS 精炼处理时,排渣后浸罩插入钢水内将空气与钢液面隔离,底吹氩气通过钢水进入浸罩,并通过上烟气管道逐步将罩内空气排出,浸罩内形成氩气氛。对于 CAS 精炼,在底吹氩产生的气泡内和钢液面上方的浸罩内均形成氩"真空"反应区。由于 CAS 精炼可以防止空气进入,没有二次氧化问题,因此可以通过钢包底部吹入大量的氩气,从而可以充分发挥氩气泡冶金的精炼功能,如图 2-25 所示。

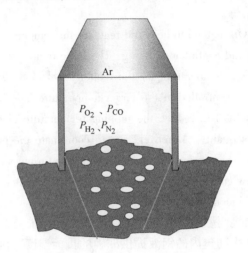

Figure 2-25 Schematic diagram of CAS-OB bottom blowing process
图 2-25 CAS-OB 底吹过程示意图

In the process of CAS-OB bottom blowing argon, due to the rising liquid flow generated by bottom blowing argon, strong stirring energy is generated in the immersion hood. Under the action of liquid steel flow, refining slag is drawn into the deeper position in the liquid steel, greatly increase the contact area between refining slag and liquid steel in the reaction zone, and improve the refining effect of slag steel. In the process of ladle bottom argon blowing, a large number of argon bubbles are produced in the molten steel. The purity of argon in the molten steel is over 99.98%. For a ladle of 50~60t, its molten steel depth is about 2.4m. At the bottom of the ladle, the internal pressure of the argon bubble is $2.68 kg/cm^2$, and the pressure of other gases except argon should be less than $0.0536 kg/cm^2$ (5.3kPa). The residual gas in argon is mainly nitrogen, so the partial pressure of nitrogen in the argon bubble should be less than $0.0536 kg/cm^2$ (5.3kPa) at the bottom of the ladle. For CO and H_2, the partial pressure is almost zero. At the steel level, the total pressure in the argon bubble is close to atmospheric pressure. The state of each gas in the argon bubble is close to that in the RH light place (4~6kPa).

在 CAS-OB 底吹氩过程中,由于底吹氩产生上升液流的作用,浸罩内产生较强的搅动能,精炼渣在钢液流的作用下,卷入钢液内比较深的位置,从而大大增加了反应区的精炼渣与钢液接触面积,提高了渣钢精炼效果。在钢包底吹氩处理过程中,钢液内产生大量的氩气泡上浮。由于吹入钢水内氩气的纯洁度达到 99.98%以上。对于 50~60t 钢包的钢水深度为 2.4m 左右，在包底时，氩气泡内部压力为 2.68kg/cm²，除氩气以外的其他气体压力应小于 0.0536kg/cm²（5.3kPa）。氩气内的残余气体主要是氮气，因此在钢包底部氩气泡内的氮气分压应小于 0.0536kg/cm²（5.3kPa），对于 CO、H₂ 来说，几乎是真空状态，其分压接近 0。在钢液面处时，其氩气泡内的总压力接近大气压。氩气泡内各气体所处状态与 RH 轻处理（4~6kPa）的条件接近。

The bottom vent plug is used to blow argon for stirring, and before the closed isolation cover enters the molten steel, the argon is used to blow away the scum on the steel liquid level from the bottom. With the rise of a large amount of argon, there is no slag in the cover, and the cover is filled with argon to form an oxygen free zone. The scum in the ladle is separated by the upper closed conical isolation hood, which provides the necessary buffer and reaction space for the oxygen flow impacting the molten steel and the oxidation reaction of aluminum and silicon. At the same time, the floating stirring argon is contained to provide argon protection space so as to improve the yield of the added aluminum, silicon and other alloy elements when the composition is fine adjusted. On the premise of preventing the secondary oxidation of the molten steel and ensuring that the temperature loss of the molten steel is small, the proper high-volume stirring of the molten steel in the early stage of refining is conducive to the purification and refining efficiency of the molten steel. CAS-OB refining equipment inserts the closed immersion hood under the liquid level in the ladle, so that the exposed molten steel is separated from the air and protected by the argon in the hood. Therefore, CAS-OB refining process can realize bottom blowing and agitation of atmospheric volume, and play the role of efficient refining.

采用包底透气塞吹氩搅拌，在封闭隔离罩进入钢水前，用氩气从底部吹开钢液面上浮渣。随着大量氩气的上浮，使得罩内无渣，并在罩内充满氩气形成无氧区。采用上部封闭式锥形隔离罩隔开包内浮渣，为氧气流冲击钢液及铝、硅氧化反应提供必需的缓冲和反应空间。同时容纳上浮的搅拌氩气，提供氩气保护空间，从而在微调成分时，提高铝、硅等合金元素的收得率。在防止钢水二次氧化，确保钢水温度损失不大的前提下，在精炼前期对钢水进行适当的大气量搅拌，有利于钢水的净化和精炼效率。由于 CAS-OB 精炼装备将密闭浸罩插入钢包液面以下，裸露的钢水与空气隔开，并受到罩内的氩气保护。所以，CAS-OB 精炼过程中可以实现大气量底吹搅拌，发挥高效精炼的作用。

Effect of CAS-OB on Inclusions in Steel
CAS-OB 对钢中夹杂物的影响

The content of inclusions in the steel after oxygen blowing is basically the same as that before treatment. But the content of inclusions after CAS treatment is very low. The total oxygen content in the steel increases slightly after OB treatment. It can be concluded that [Si] and [Mn] have a certain oxidation tendency, [S] and [P] have a little recovery, and [C] and [N] have no big fluctuation.

在吹氧后，钢中夹杂物含量基本与处理前持平，但经 CAS 处理后夹杂物含量非常低，钢中总氧含量在 OB 处理后略有上升。CAS-OB 对钢水成分的影响归纳为：[Si] 和 [Mn] 有一定的氧化倾向，[S] 和 [P] 稍有回复，[C] 和 [N] 没有出现大的波动。

2.7.4.2 CAS-OB Refining Process System

2.7.4.2 CAS-OB 精炼工艺制度

CAS-OB refining has the advantages of fast heating speed, high thermal efficiency, less equipment investment, low operation cost, and simple control process. CAS treatment after oxygen blowing ensures the floatation and rapid removal of a large number of small Al_2O_3 inclusions after oxygen blowing, and ensures that the quality of molten steel is not affected by argon blowing. The main process systems include:

CAS-OB 精炼升温速度快,热效率高,设备投资少,操作成本低,控制过程简便、快速、准确。吹氧后再进行 CAS 处理,保证了吹氧后大量细小 Al_2O_3 夹杂的上浮和快速去除,同时也保证了钢水质量不受吹氩的影响。其主要工艺制度包括:

(1) Temperature rise and terminal temperature control of argon blowing: In the process of oxygen blowing, adding aluminum shot continuously and controlling Al/O_2 ratio are the key technologies to avoid burning loss of C, Si, Mn and other elements in steel and to control acid soluble aluminum content in steel. It belongs to bulk heating and its thermal efficiency is higher than 90%. Generally, the temperature of each ton of molten steel is increased by 1℃, and the aluminum consumption is 350~450g. The heating rate is fast, and the whole CAS-OB treatment cycle is 11~16min, which can match the converter production rhythm. The main factors affecting the heating rate of CAS-OB are as follows:

(1) 吹氩升温和终点温度控制:在吹氧过程中,连续加入铝丸,控制 Al/O_2 比是避免钢中 C、Si、Mn 等元素烧损和控制钢中酸溶铝含量的关键技术。溶解铝氧化升温工艺属于体相加热,热效率高于 90%。通常每吨钢水升温 1℃,耗铝量为 350~450g。升温速度快,整个 CAS-OB 处理周期为 11~16min,可与转炉生产节奏相匹配。影响 CAS-OB 升温速度的主要因素为:

1) Al content: High Al excess coefficient (the ratio of chemical equivalent supplied by Al/O_2) can greatly improve the heating rate and heating efficiency.

1) Al 含量:高的 Al 过量系数(指 Al/O_2 供入的化学当量的比值)可以大大提高升温速率和加热效率。

2) Oxygen blowing: High oxygen supply intensity is the guarantee of rapid temperature rise, but excessive oxygen blowing may cause molten steel oxidation. Generally, it can be solved by adjusting the amount of argon blown at the bottom, increasing the amount of argon blown at the bottom, speeding up the circulation of molten steel, and making the molten steel difficult to oxidize due to the reaction composition of aluminum oxide. In addition, under the condition of strong agitation, the heating agent burns efficiently, the heat transfer condition is good, and the heating rate is increased.

2) 吹氧量:高的供氧强度是快速升温的保障,但过量吹氧可能造成钢水氧化。一般可通过调整底吹氩量解决。底吹氩量增加,钢水循环速度加快,氧铝反应成分,钢水不易氧化。另外在强搅拌条件下,发热剂高效燃烧,传热条件良好,升温速度提高。

3) Lance position of oxygen lance: The lower lance position increases the impact strength of the jet, resulting in a larger fire point area surrounded by molten steel, with less heat loss. But the lance position is too low, the melting loss of oxygen lance intensified, the control of lance position difficult, and the temperature raising effect will be affected. Within the range of 50~800mm controllable lance position, the lance position of 200~500mm is relatively ideal.

3) 氧枪枪位：低枪位增加了射流的冲击强度，从而造成更大的、同时被钢水包围的火点区，使得热损失少。但由于枪位过低，氧枪熔损加剧，枪位控制困难，提温效果也受到影响。在50~800mm可控枪位范围内，200~500mm枪位是比较理想的。

4) Ladle capacity, type of heating agent, original molten steel composition, and size of isolation cover.

4) 钢包容量、发热剂种类、原始钢水成分和隔离罩尺寸。

(2) Ar blowing process and inclusion removal: When the temperature is increased by adding aluminum, a large amount of Al_2O_3 inclusions are formed by aluminum oxidation, which may increase the aluminum content in the steel. During the heating process, the ratio of Al/O_2 and the stirring intensity are precisely controlled. After the heating, the Ar stirring is guaranteed for a certain time to promote the inclusion floating.

(2) 吹Ar工艺与夹杂物去除：采用加铝升温，铝氧化生成大量Al_2O_3夹杂，并可能使钢中铝含量升高。因此在加热过程中需要精确控制Al/O_2比及搅拌强度，升温后保证一定时间的吹氩搅拌，促进夹杂物上浮。

(3) Alloy fine tuning process: During tapping, the composition of molten steel shall be fine adjusted, and the steel sample shall be taken for rapid analysis. According to the results of chemical analysis, the alloy shall be added for the final adjustment of the composition of molten steel in CAS treatment to achieve accurate control of ingredients.

(3) 合金微调工艺：出钢时，对钢水成分进行精调，并取钢水样进行快速分析。根据化学分析结果，在CAS处理中补加合金进行钢水成分的最终调整，实现成分准确控制。

CAS-OB process: Converter tapping→Ladle reaching the processing position→Connected to argon blowing pipe → Temperature measurement (oxygen determination) sampling → Argon stirring→Dip tube down→Composition and temperature adjustment→Temperature measurement and sampling→Tip adjustment of composition and temperature→Dip tube rising→Argon blowing stop→Temperature measurement and sampling→Pull out argon pipe→Ladle leaving.

CAS-OB工艺流程：转炉出钢→钢包到达处理位→接吹氩管→测温（定氧）取样→吹氩搅拌→浸渍管下降→成分、温度调整→测温取样→成分、温度微调→浸渍管上升→吹氩停止→测温取样→拔吹氩管→钢包离开。

2.7.4.3 CAS-OB Refining Equipment

2.7.4.3 CAS-OB 精炼设备

CAS-OB equipment mainly includes: bottom blowing argon system, alloy weighing and adding system, oxygen gun lifting and blowing oxygen system, immersion hood and its lifting system, insulation material adding system, automatic temperature measurement and sampling device, pneumatic sample sending and spectrum analysis system, and dust removal system. The equipment is shown in Figure 2-26.

CAS-OB 设备主要包括：底吹氩系统，合金称量与加入系统，氧枪升降与吹氧系统，浸渍罩及其升降系统，保温材料加入系统，自动测温取样装置，风动送样及光谱分析系统，除尘系统。其设备示意图如图 2-26 所示。

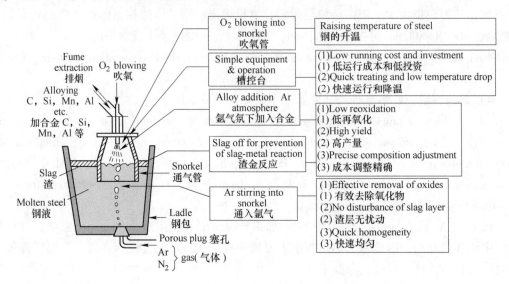

Figure 2-26 Schematic diagram of CAB-OB equipment
图 2-26 CAB-OB 设备示意图

Isolation Cover (Immersion Cover)
隔离罩（浸渍罩）

The function of the isolation cover includes: isolating the dross, forming the dross free liquid surface, and providing the space for adding fine-tuning alloy to improve the recovery rate of alloy elements; accommodating the floating stirring argon, and providing the argon protection space; providing the reaction space for the aluminum oxygen reaction heating steel liquid; collecting and exhausting the flue gas (smoke cover function). CAS-OB isolation hood is conical, which is welded by steel plate and divided into upper and lower parts. The upper cover is lined with refractory materials, and the lower cover is lined with refractory materials both inside and outside, so that it can be immersed in the molten steel. Generally, the immersion depth is 100~200mm. Refractories are high alumina amorphous materials with $w(Al_2O_3) > 70\%$, with a service life of 65~100 times, as shown in Figure 2-27.

隔离罩的作用包括：隔离浮渣，形成无渣液面，并提供加入微调合金空间，提高合金元素的收得率；容纳上浮的搅拌氩气，提供氩气保护空间；为铝氧反应加热钢液提供反应空间；收集和排出烟气（烟罩作用）。CAS-OB 隔离罩为锥形体，由钢板焊成，分上下两部分。上罩体内衬耐火材料，下罩体内外均衬以耐火材料，以便浸入钢液内部，通常浸入深度为 100~200mm。耐火材料为高铝质不定型材料，$w(Al_2O_3) > 70\%$，使用寿命为 65~

100次。隔离罩设备示意图如图2-27所示。

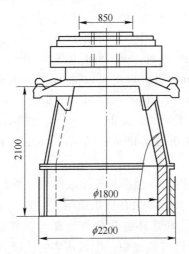

Figure 2-27　CAS-OB isolation hood
图2-27　CAS-OB隔离罩设备示意图

The position of the isolation cover in the ladle shall be able to cover all the floating argon bubbles basically and keep a proper distance from the ladle wall. According to the experience of Nippon Steel, the distance from the inner wall of the package to the outer wall of the isolation cover is 170mm.

隔离罩在钢包内的位置应能基本笼罩住全部上浮氩气泡，并与包壁保持适当距离。据新日铁生产经验，自包内壁至隔离罩外壁的距离为170mm。

The inner diameter of the isolation hood shall be determined by：
隔离罩的内径的计算公式为：

$$D = 2\tan(\theta/2)h + d \tag{2-13}$$

Where　D——the inner diameter of isolation cover in m；
　　　　h——the liquid steel depth in m；
　　　　θ——the expansion angle of argon bubble in 20°~25°；
　　　　d——the diameter of argon blowing brick in m.
式中　D——隔离罩内径，m；
　　　h——钢液深度，m；
　　　θ——氩气泡扩张角，取20°~25°；
　　　d——吹氩砖直径，m。

The depth of isolation (100~200mm) cover 隔离罩深度 = Clearance height 净空高度 + Slag layer thickness 渣层厚度 + The depth of cover inserted into molten steel 罩插入钢液中的深度

$$\tag{2-14}$$

Oxygen Lance
氧枪

In CAS-OB refining, inert gas is often used to surround the oxygen flow stream. The common

· 161 ·

oxygen lance is of consumption type, which is composed of double-layer stainless steel tubes. The casing is coated with high alumina refractory [$w(Al_2O_3) \geqslant 90\%$], and the casing clearance is generally 2~3mm. Oxygen is blown in the central pipe and argon is blown in the casing ring seam for cooling. The argon volume accounts for about 10% of the oxygen volume. This spray gun makes the oxygen stream surrounded by inert atmosphere, forming a concentrated oxygen blowing point, and forming a low oxygen partial pressure area at a large steel liquid level, thus inhibiting the oxidation of the steel liquid. The ratio of the outer pipe pressure to the inner pipe pressure can be controlled between 1.2 and 3.0. The burning rate of the oxygen gun is about 50mm/time, and its service life is 20~30 times.

CAS-OB 精炼中多采用惰性气体包围氧气流股的双套管顶吹氧枪。常见的吹氧枪为消耗型，用双层不锈钢管组成。套管外涂有高铝质耐火材料 [$w(Al_2O_3) \geqslant 90\%$]，套管间隙一般为 2~3mm。中心管吹氧，套管环缝吹氩气冷却，氩气量大约占氧气量的 10% 左右。这种喷枪使氧气流股包围在惰性气氛当中，形成集中的吹氧点，大的钢液面则形成低氧分压区，从而抑制钢液的氧化。操作中控制外管压力和内管压力的比值在 1.2~3.0 之间，氧枪的烧损速度大约为 50mm/次，寿命为 20~30 次。

In order to improve the heating rate, the lower section of the top oxygen lance is placed in the foaming steel liquid formed by bottom blowing gas and top blowing oxygen in the isolation hood for oxygen blowing. The most reasonable method is to bury the gun head in the foaming molten steel. Obviously, the inert gas blown at the bottom is not enough to make the molten steel foaming. A large number of CO and O_2 gases are generated by the oxidation of carbon in the steel. Blowing oxygen in the foaming molten steel can achieve high thermal efficiency and rapid temperature rise. Because the heat transfer to the molten steel is the fastest at this time, and the gun position control is extremely important here. In the process of oxygen blowing, a certain amount of carbon should be consumed. The solution is to tap the converter with the upper deviation of carbon, or add some carbon after OB progress.

为提高升温速度，将顶吹氧枪下段置于由底吹气体和顶吹氧在隔离罩内形成的发泡钢液中进行吹氧。最合理的方法是把枪头埋入发泡的钢液中。显然，底吹的惰性气体不足以使钢液发泡，因此要借助钢中碳的氧化生成大量的 CO 和 O_2 气体。在起泡的钢液中吹氧可以实现高热效率，以及快速升温。因为此时热量向钢液的传递速度最快，在这里枪位的控制极为重要。在这种吹氧过程中，要消耗一定量的碳，其解决方法是转炉以碳的上偏差出钢，或 OB 后补加部分碳。

2.7.4.4 CAS-OB Refining Effect and Evaluation

2.7.4.4 CAS-OB 精炼效果及评价

After CAS-OB refining, the liquid steel can increase the gas-liquid ratio (the ratio of argon to liquid steel) in the ladle, increase the reaction amount between argon bubble and liquid steel, promote the reaction between [O] with [C], [Al] and [Si] in the molten steel, greatly reduce the dissolved oxygen content in the molten steel, and improve the probability of small inclusions colliding, polymerizing and floating upward. Air mixing is beneficial to the reaction of slag and steel. The refining effects of CAS-OB are as follows:

钢液经过 CAS-OB 精炼处理后提高钢包内气液比（氩气与钢水量比），增大氩气泡与钢水的反应量；促进钢水内 [O] 与 [C]、[Al]、[Si] 等的反应，大大降低钢水中溶解氧量；提高细小夹杂物碰撞聚合上浮、气泡吸附夹杂物上浮的概率。同时，大气量搅拌有利于渣钢混合促进渣钢反应。CAS-OB 的精炼效果包括：

(1) The total oxygen content in the steel can be reduced from 0.010% to less than 0.0040% compared with that in the steel treated by conventional argon blowing and stirring.

(1) 与常规吹氩搅拌处理的钢液相比，钢中总氧量可由 0.010% 下降到 0.0040% 以下。

(2) The large inclusions above 40 m can be reduced by 80%, and the inclusions of 20~40m can be reduced by 1/2~1/3, which greatly improves the quality of steel.

(2) 40m 以上的大型夹杂物可减少 80%；20~40m 的夹杂物可减少 1/2~1/3，大大改善钢的质量。

(3) The yield of deoxidizing elements is increased. Due to the increase of yield, the amount of acid soluble aluminum in steel can be multiplied with the same amount of aluminum added. If some slag remains in the cover, the yield of the alloy will decrease, andthe operation is not stable. The practice shows that when the slag layer is too thick, the capacity of slag removal and slag separation of the isolation hood can not be fully utilized.

(3) 提高脱氧元素的收得率。由于收得率的提高，在同样的加铝量中，钢中酸溶铝量可成倍增加。若有部分渣残留在罩内，合金收得率就会下降，且操作不稳定。实践表明，渣层过厚时，隔离罩的排渣、隔渣能力得不到充分利用。

The comparison between CAS-OB and other refining technologies, and the comparison between CAS with LF, RH and VD refining effects are shown in Table 2-8. CAS has the same advantages as other refining devices in composition regulation, deoxidation and inclusion removal, but there is still a certain gap in desulfurization, degassing and slag refining. However, CAS has great advantages in processing time, preventing secondary oxidation of molten steel, processing cost, preventing carbon and nitrogen increase, etc.

CAS-OB 与其他精炼技术以及 CAS 与 LF、RH、VD 精炼效果对比见表 2-8。CAS 在成分调节、脱氧、夹杂物去除方面与其他精炼装置有同等优点，而在脱硫、脱气、渣精炼方面还有一定的差距。然而在处理时间，防止钢水二次氧化，处理成本，防止增碳增氮等方面具有很大的优越性。

Table 2-8　Comparison between CAS-OB and other refining technologies
表 2-8　CAS-OB 与其他精炼技术对比

Project 项目	CAS	LF	RH	VD
Composition adjustment 成分调节	Good 好	Good 好	Good 好	Good 好

Continued Table 2-8

Project 项目	CAS	LF	RH	VD
Deoxidation 脱氧	Good 好	Good 好	Good 好	Good 好
Desulphurization 脱硫	Bad 差	Good 好	Bad 差	Good 好
Processing time 处理时间	Quick 快	Slow 慢	Slow 慢	Slow 慢
Mild steel 低碳钢	No carburizing 不增碳	Add 增碳	Deep decarburization 深脱碳	Deep decarburization 深脱碳
Degassing 脱气	No 无	No 无	Yes 有	Yes 有
Slag refining 渣精炼	No 无	Good 好	No 无	Good 好
Inclusion removal capacity 夹杂物去除能力	Good 好	Good 好	General 一般	Good 好
Secondary oxidation 防二次氧化	Good 好	General 一般	Good 好	Good 好
Processing cost 处理成本	Low 低	High 高	High 高	High 高
Nitrogen increasing 增氮	Less 少	More 较高	Less 少	Less 少

CAS-OB refining has been gradually used in China since the 1980s, and there are some problems in the process of application, including that: most of the converters in China are small and medium-sized, the production process and workshop have been completed, and the process of BOF-CC has a relatively fast pace. Therefore, the original CAS technology can not meet the requirements of many steel plants in terms of production process layout, equipment construction, investment cost, production process, etc. If the scale and capacity of steelmaking system is too small, CAS-OB should not be used for refining. In addition, the problems and solutions in CAS-OB refining process are summarized as follows:

我国自20世纪80年代就开始逐步使用CAS-OB精炼,使用过程也出现了一些工艺条件不匹配的问题,主要包括:我国大多为中小转炉,生产流程和车间厂房均已建设完毕,转炉—连铸生产节奏比较快,因此原CAS技术在生产流程中的布置方式、设备建设、投资成本、生产工艺等方面满足不了我国许多钢厂现场条件与生产需要的要求。当炼钢系统规模及容量过小时,不宜使用CAS-OB精炼。CAS-OB精炼使用过程中出现的问题及解决方法归纳如下:

(1) Peroxide of molten steel and loss of beneficial elements: The solution is to increase the argon flow rate of bottom blowing, control the oxygen flow rate of top blowing and increase the ex-

cess coefficient of aluminum.

（1）钢水过氧化和有益元素的损失：其解决方法为增加底吹氩流量、控制顶吹氧流量以及增加铝的过量系数。

（2）Burning loss of oxygen lance: It is recommended to use high alumina refractory lined with stainless steel pipe.

（2）氧枪烧损：建议使用不锈钢钢管外衬高铝质耐火材料。

（3）Service life of isolation cover: Control the proper gun position to prevent excessive splashing, and surround the fire point area in the molten steel as much as possible to reduce the direct radiation to the cover lining. In addition, add some aluminum before blowing oxygen to reduce the large amount of FeO enrichment on the surface.

（3）隔离罩寿命：控制合适的枪位防止喷溅过度，尽可能将火点区包围在钢水中减少对罩衬的直接辐射。另外，在吹氧前先加入部分铝，减少表面 FeO 的大量富集。

2.7.4.5 CAS-OB Refining Process and Equipment Innovation

2.7.4.5 CAS-OB 精炼工艺及设备创新

Online CAS-OB refining
在线 CAS-OB 精炼

CAS-OB refining units applied in foreign countries and newly built in China are all arranged offline, and the refining station is built outside the tapping track line of converter production process. The advantage is that the operation of tapping in the next furnace will not be affected during CAS-OB refining. However, the disadvantage is that the number of ladle dispatching links and the auxiliary time are increased, and the effective time of refining is reduced. For the large converter, the smelting time is long and the site environment is spacious. The construction of offline CAS-OB can meet the requirements from both production time and workshop space.

在国外应用的 CAS-OB 精炼装置与国内新建的 CAS-OB 精炼装置均采用离线布置的方式，即精炼站建立在与转炉生产流程的出钢轨道线外的位置。其优点是在 CAS-OB 精炼处理时，不影响下一炉出钢的操作；所带来的缺点是钢包调运环节增多，增加了辅助时间，减少了精炼的有效时间。从生产时间和车间空间来说，冶炼时间长，现场环境宽敞，以及建设离线 CAS-OB 都能满足大转炉要求。

In China, the medium and small-sized converters adopt the fast-paced production mode. The converter smelting and casting cycle are relatively short, and plus the ladle transportation time. The time left for CAS-OB refining treatment is shorter, which will greatly affect the refining effect. In order to adapt to the actual situation of our country, CAS-OB refining station can be built online. CAS-OB can be built at the original position of online argon blowing station, which can save 5~8min of ladle dispatching time, and solve a series of problems such as online replacement of immersion hood, rapid refining, online monitoring of ladle car operation, online rapid analysis, rapid detection of slag surface, etc. It can also improve the refining efficiency of CAS-OB and meet the needs of medium and small converter for fast-paced production.

我国中小转炉采用快节奏的生产方式,转炉冶炼与连铸浇注周期都比较短,再加上钢包调运的时间,留给 CAS-OB 精炼处理的时间就更短,这样就会大大影响精炼效果。为了适应我国实际情况,可建设在线 CAS-OB 精炼站,将 CAS-OB 建设在原在线吹氩站位置处,可节约多余的钢包调运时间 5~8min,并解决了浸罩在线更换、快速精炼、钢包车运行在线监控、在线快速分析、渣面快速检测等一系列问题,同时提高了 CAS-OB 精炼效率,满足中小转炉快节奏生产的需要。

CAS-OB Equipment Structure Optimization Design
CAS-OB 设备结构优化设计

CAS-OB equipment imported from Japan has complex structure, requiring high workshop space, large floor area, large structure of mechanical equipment and large investment cost. On the premise of not affecting CAS-OB refining capacity, the research and development of domestic independent intellectual property rights of CAS-OB equipment were carried out. The cutting mode, immersion hood structure and lifting mechanism, oxygen gun structure and lifting system, steel liquid level detection system, overall structure of equipment and computer control system of equipment electrical meter were comprehensively reformed. The new CAS-OB refining device has the characteristics of small floor area, low space height, simple equipment structure, convenient operation, low equipment failure rate and less investment, which is suitable for the needs of medium and small converter equipped with on-line CAS-OB refining in China.

日本引进的 CAS-OB 装备结构比较复杂,要求厂房空间高,占地面积大,机械设备结构比较庞大,投资成本大。在不影响 CAS-OB 精炼能力的前提下,对其装备进行了国产化自主知识产权化的研究开发,对下料方式、浸罩结构、升降机构、氧枪结构、升降系统、钢液面检测系统、设备的总体结构和设备电气仪表计算机控制系统进行了全面改革。新型 CAS-OB 精炼装置具有占地面积少、空间高度低、设备结构精简、操作方便、设备故障率低、投资少等特点,适合我国中小转炉配备在线 CAS-OB 精炼的需要。

Development of CAS-OB Multifunctional Refining Process
CAS-OB 多功能精炼工艺开发

The main refining functions of conventional CAS-OB are: regulating the composition of molten steel, heating up the molten steel, preventing the secondary oxidation of molten steel, purifying the molten steel, etc. In order to better meet the production needs, on the basis of the original function, the top slag desulfurization technology of CAS, which is improved by slag inclusion and stirred by large volume in CAS immersion hood, is developed, and the desulfurization effect is 20%~45%.

According to the actual situation that many steel plants in China are equipped with slab caster and billet caster, the aluminum killed steel (mainly slab caster) and silicon killed steel (mainly billet continous caster) are developed, which can effectively prevent the clogging of the nozzle.

常规 CAS-OB 主要的精炼功能有:钢水成分调节、钢水升温、防止钢水二次氧化、钢水净化等。为了更好地满足生产需要,在原功能的基础上,开发了通过包渣改质、CAS 浸罩内大气量搅拌的 CAS 顶渣脱硫技术,脱硫效果达到 20%~45%;针对我国许多钢厂炼钢比配备板坯连铸机又配备小方坯连铸机的实际情况,开发了铝镇静钢(主要是板坯连铸)

和硅镇静钢（主要是小方坯连铸）的钢水升温工艺，有效防止了连铸水口可能出现堵塞的现象等。

Optimize Oxygen Blowing System and Equipment

优化吹氧制度及设备

Development of double sleeve oxygen lance is a double sleeve oxygen lance with inert gas surrounding the oxygen flow strand developed in Bafan Plant of Japan, which makes the oxygen flow strand surrounded by inert atmosphere, forms a concentrated oxygen blowing point and a low oxygen partial pressure area on the steel surface, and inhibits the oxidation of the steel. At present, this type of oxygen lance has been used in some steel plants at home and abroad.

开发双套管氧枪是日本八幡厂开发的一种有惰性气体包围氧气流股的双套管喷枪，氧气流股包围在惰性气氛当中，并形成集中的吹氧点，在钢面形成低氧分压区，抑制了钢水的氧化。目前该型号氧枪已经在国内外的一些钢厂试用。

Add aluminum to steel before oxygen blowing. When heating agent and oxygen are supplied to the steel surface at the same time, the steel and heating agent in the isolation cover oxidize at the same time to form a FeO rich oxide slag layer. The slag layer reacts with the aluminum added continuously, and the heat generated can only be absorbed by the steel on the surface, and it cannot be transmitted to the lower steel. The oxide layer also causes the steel pollution, decarburization and refractory loss. Before oxygen blowing, 0.5kg/t steel aluminum is added to the surface of molten steel. The heating agent forms a melting layer of heating agent on the surface of molten steel within 10~30s. Oxygen blowing is carried out immediately and heating agent is put in at the same time, so that the aluminum can be oxidized preferentially, forming a slag layer with good fluidity, no pollution to molten steel and no erosion of the isolation cover. And the generated heat can be transferred to the interior of molten steel smoothly.

吹氧前向钢中添加铝。当发热剂与氧气同时供向钢水表面时，隔离罩内的钢水和发热剂同时氧化形成富含 FeO 的氧化渣层。该渣层与连续添加的铝发生反应，产生的热量只能被表面的钢水吸收，无法传至下层钢水，氧化层还会造成钢水污染、脱碳和耐材损耗。在吹氧前向钢水表面添加 0.5kg/t 钢铝，发热剂在 10~30s 内在钢水面形成发热剂熔融层，立即吹氧，同时投入发热剂。这样铝便可以优先氧化，形成流动性良好的渣层，对钢水无污染，也不会侵蚀隔离罩，同时使产生的热量顺利向钢水内部传递。

The insertion depth of the isolation hood is changed by constant bottom blowing argon volume. The insertion depth is 200mm for CAS treatment and 400mm for OB treatment.

采用恒定的底吹氩量，改变隔离罩插入深度，合理控制 CAS 处理时插入深度 200mm，OB 处理时则为 400mm。

2.7.5 IR-UT

2.7.5 IR-UT

IR-UT (Injection Refining with Temperature Raising Capability) is the ability to increase the temperature of the injection refining. This is another new refining method developed by Sumitomo Metal Industry Co., Ltd. in 1986 after CAS-OB method, which is called IR-UT ladle metallurgy station. The technology is equipped with two immersion spray guns for blowing oxygen and stirring (or spraying powder) with argon blowing. The argon blowing mode is changed from bottom blowing to top blowing. After heating the liquid steel with oxygen blowing, the operation of powder spraying, desulfurization and inclusion shape adjustment is also a method of CAS refining.

IR-UT 是指带有增加温度能力的喷吹精炼。这是继 CAS-OB 法后，日本住友金属工业株式会社于 1986 年开发的又一项新的炉外精炼方法，称之为 IR-UT 钢包冶金站。该技术设置了吹氧和吹氩搅拌（或喷粉）用的两支浸入式喷枪，吹氩方式由底吹改为顶吹，对钢液吹氧加热后进行喷粉脱硫和夹杂物形态的调整操作，也属于 CAS 精炼的一种方法。

The IR-UT method can effectively stir the liquid steel and make the composition and temperature of the liquid steel uniform. The sulfur can be reduced to a low level of 0.001%~0.010% by using the powder spray tank. Although aluminum is added during tapping and oxygen blowing, which increases the alumina content in steel, the alumina content in most furnaces is lower than that of unheated furnaces after argon blowing and stirring. The purity of liquid steel is enough to meet the requirements of continuous casting. The processing time of IR-UT is about 20min, and the heating time is about 5min. IR-UT method and refining effect are the same as CAS-OB method, but the desulfurization capacity is improved. Compared with CAS-OB refining, its advantages are as follows:

IR-UT 法取消了包底的多孔透气砖，改底吹氩为顶吹氩，可以有力地搅拌钢液，促使钢液成分和温度均匀。使用喷粉罐脱硫后，硫可以降到 0.001%~0.010% 的低水平。虽然出钢时加铝，吹氧时还加铝，增加了钢中的氧化铝含量，但是吹氩搅拌后，大多数炉次的钢中氧化铝含量低于不加热炉次的水平。钢液纯净度足以满足连铸要求。IR-UT 处理时间一般为 20min 左右，其中加热时间约为 5min。IR-UT 法和精炼效果与 CAS-OB 法相同，但脱硫能力提高。对比 CAS-OB 精炼其优点如下：

(1) It is simple and rapid to adjust the composition and temperature of molten steel uniformly.

(1) 均匀调节钢液成分和温度简单迅速。

(2) The alloy yield is improved.

(2) 提高合金收得率。

(3) The accuracy of carbon and manganese can be $15 \times 10^{-3}\%$ and $30 \times 10^{-2}\%$ respectively.

(3) 微调成分，碳的精确度可达 $15 \times 10^{-3}\%$；锰的精确度可达 $30 \times 10^{-2}\%$。

(4) The inclusion content is obviously reduced and the purity of molten steel is high.

(4) 夹杂物含量明显减少，钢液纯净度高。

(5) The equipment is simple, no complex vacuum equipment is needed, capital construction investment is saved, and the cost is low. The investment of CAS-OB method is 1/8 of DH method, and the operating cost is 1/5 of DH method.

(5) 设备简单，无须复杂的真空设备，基建投资省，成本低。CAS-OB 法投资是 DH 法的 1/8，操作费用为其 1/5。

(6) In IR-UT method, argon blowing at the bottom is eliminated, and there is no steel leakage from the permeable brick.

(6) IR-UT 法取消了底吹氩气，不存在从透气砖处漏钢问题。

(7) It can be used not only in large ladle, but also in small ladle, and it has more extensive adaptability.

(7) 不仅适用于大钢包，也可以用于小钢包，具有更广泛的适应性。

The IR-UT ladle metallurgy station is composed of the following equipment: ladle cover and lifting mechanism with isolation cover, top blowing argon mixing gun and lifting machinery, oxygen gun and lifting mechanism, blowing (CaO powder, CA-Si powder) tank and hose, alloy weighing and adding system, sampling and temperature measuring device, connecting pipe lifting winch, scrap adding device, wire feeding system (optional equipment), and powder spray tank for injection (optional equipment) in lime powder or Ca-Si powder. IR-UT method is also equipped with direct reading spectrometer, pneumatic sample delivery, dust removal system, etc., as shown in the Figure 2-28.

IR-UT 法钢包冶金站由下列设备组成：带隔离罩的钢包盖和提升机构，顶吹氩搅拌枪和提升机械，吹氧枪和提升机构，喷吹（CaO 粉，Ca-Si 粉）罐和软管，合金料称量和加入系统，取样和测温装置，连通管升降卷扬机，加废钢装置，喂线系统（任选设备），以及石灰粉或 Ca-Si 粉喷吹用的喷粉罐（任选设备）。IR-UT 法还配有直读光谱仪，风动送样，除尘系统等，如图 2-28 所示。

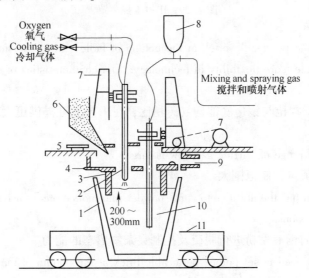

Figure 2-28 IR-TU equipment composition
图 2-28 IR-UT 设备构成

1—Ladle; 2—Oxygen lance; 3—Isolation cover; 4—Cover; 5—Platform; 6—Alloy weighing unit; 7—Lifting device; 8—Spray tank; 9—Exhaust port; 10—Mixing gun; 11—Ladle car
1—钢包；2—吹氧枪；3—隔离罩；4—包盖；5—平台；6—合金称量单位；7—升降装置；8—喷射罐；9—排气口；10—搅拌枪；11—钢包车

The isolation cover of IR-UT is cylindrical, and the top surface has flange to cover the tank mouth. The upper part is connected with the flat cover, which can cover the opening of the ladle and support it on the edge when it drops to the lowest position. The immersion depth changes with the fluctuation of the steel liquid level. The IR-UT isolation cover is arranged concentrically with the ladle due to the top gun mixing, as shown in Figure 2-29.

IR-UT 的隔离罩为筒形，顶面有凸缘可盖住罐口。上部与平盖连在一起，下降到最低位时能盖住包口并支撑在包沿上，浸入深度随钢液面波动而变化。IR-UT 隔离罩因采用顶枪搅拌而采取与钢包同心布置，如图 2-29 所示。

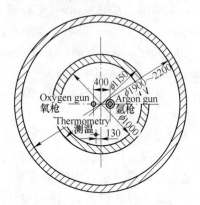

Figure 2-29　IR-UT isolation hood
图 2-29　IR-UT 隔离罩

Compared with the permeable brick at the bottom of ladle, the immersion top blowing argon stirring gun can provide more flexibility in technology. The characteristics of spray gun are as follows:

浸入式顶吹氩搅拌枪与钢包底部透气砖相比，在工艺上可提供更大的灵活性。该喷枪具有以下特点：

(1) The control range of stirring gas flow is large.
(1) 搅拌气体流量控制范围大。
(2) The top gun has the abilities to spray powder into molten steel for desulfuration and control the shape of inclusions.
(2) 顶枪具有向钢液喷粉进行脱硫及控制夹杂物形态的能力。
(3) There is no need to set porous permeable brick at the bottom of the bag.
(3) 无须设置包底多孔透气砖。
(4) It can reheat the whole ladle back from the continuous caster. But it is difficult to heat the ladle with permeable brick, because the low temperature of the ladle bottom will cause the solidification of the molten steel near the permeable brick and hinder the air flow.
(4) 能对连铸机返回来的整包钢液进行再次加热。而带有透气砖的钢包加热时有较大困难，因为包底温度低会导致透气砖近处钢液凝固，妨碍气流通入。

IR-UT ladle metallurgy station adopts the upper open type isolation hood, which has the following advantages compared with the upper closed isolation hood adopted by CAS-OB:

IR-UT 钢包冶金站采用上部敞口式隔离罩，它与 CAS-OB 采用的上部封闭的隔离罩相比具有下列优点：

(1) It can reduce the height of the whole equipment.

(1) 可使整个设备的高度降低。

(2) The feeding line can be carried out in the isolation cover to avoid the reaction with the surface slag.

(2) 喂线可在隔离罩内进行，免除与表面渣的反应。

(3) It is easy to observe and adjust various operations in the process of liquid steel treatment, such as oxygen blowing, stirring, alloying and erosion of refractory in the isolation hood.

(3) 在钢液处理过程中容易观察和调整各项操作，如吹氧、搅拌、合金化及隔离罩内耐火材料的侵蚀等。

Task 2.8　VD Refining
任务 2.8　VD 精炼

In addition to the ladle refining technology introduced before, vacuum treatment technology is widely used in the secondary refining production. The vacuum treatment of molten steel is to reduce the gas pressure in the container of molten steel treatment by means of vacuumizing before or during pouring, so as to remove the gas and non-metallic inclusions in the steel. If the liquid steel without deoxidation is treated, further decarburization can be carried out to produce low carbon steel. At present, there are many vacuum treatment methods, but they are generally divided into ladle degassing and steel flow degassing.

除了之前介绍的钢包精炼技术之外，真空处理技术广泛应用于炉外精炼生产。钢水真空处理就是在浇铸前或浇铸过程中，利用抽真空的办法来降低钢水处理容器中气体的压力，以达到去除钢中气体和非金属夹杂物的目的。若处理未脱氧的钢水时还能进一步脱碳，则能炼出低碳钢。目前，真空处理方法很多，但一般将其分为钢包脱气和钢流脱气。

Ladle degassing method is first put the ladle into the vacuum chamber, cover the vacuum chamber cover and then vacuum degassing. Steel flow degassing includes lift degassing and circulation degassing, introduced in the next tasks.

钢包脱气法是先将钢包放入真空室内，盖上真空室盖后抽真空脱气。钢流脱气法包括提升脱气法和循环脱气法，该法将在以后的任务中予以介绍。

VD (Vacuum Degassing) refining is a representative process of ladle degassing. A. Finkl & Sons in the United States combined simple ladle argon blowing with vacuum degassing to form a ladle vacuum treatment method, also known as finkl method, which was first applied in Germany in the 1950s. VD refining is a kind of vacuum treatment method that puts the primary molten steel of converter and electric furnace in a vacuum chamber and blows argon at the bottom of ladle for stirring. It can be used for decarburization, degassing, desulfuration, impurity removal, alloying and uniform temperature and composition of molten steel. It is mainly used for steel grades with strict control of liquid steel gas requirements. VD vacuum treatment relies on the whole process of argon blowing and stirring at the bottom of the ladle, in order to homogenize the composition and temperature of the molten steel, and promote the vacuum degassing, sulfur removal, composition adjustment and inclusion floating. Especially the soft argon blowing after wire feeding is an effective method to remove oxide inclusions in the steel. Compared with RH vacuum treatment process, refining strength of VD is strictly restricted by ladle clearance. Generally, it is required that the clear space of the ladle is 800~1000mm, refined under the vacuum of 13.33~266.64Pa. If the liquid steel carbon deoxidization process is carried out, the clear space of the ladle shall not be less than 900mm; and if the oxygen blowing decarbonization process is realized, the clear space of the ladle shall be 1.2~1.5m.

VD 精炼是钢包脱气法的代表工艺,是由美国芬克尔父子公司将简单的钢包吹氩与真空脱气相结合,形成一种钢包真空处理方法,也叫芬克尔法,并于20世纪50年代由德国首先应用。VD 精炼是将转炉、电炉的初炼钢水置于真空室中,同时也是钢包底部吹氩搅拌的一种真空处理法,可进行脱碳、脱气、脱硫、去除杂质、合金化和均匀钢水温度、成分等处理,主要用于对钢液气体要求严格控制的钢种。VD 真空处理依靠钢包底部全程吹氩搅拌,目的是均匀钢水的成分和温度,促进真空脱气、去硫、成分调整、夹杂物上浮,尤其是喂线后的软吹氩更是去除钢中氧化物夹杂的有效方法。与 RH 真空处理工艺相比,VD 的精炼强度受到钢包净空的严格制约。一般要求钢包净空为800~1000mm,在真空度为13.33~266.64Pa 下精炼。在进行钢液碳脱氧工艺时,钢包净空应不小于900mm;在实现吹氧脱碳工艺,则钢包净空为1.2~1.5m。

2.8.1 Production and Progress of VD Refining

2.8.1 VD 精炼工艺及流程

VD Refining is mainly coordinated with converter, electric arc furnace and LF refining. Its basic principle is based on the square root law of H and N dissolving in molten steel. When VD is vacuumized, the pressure in the vacuum chamber will be reduced, so that [H] and [N] will be reduced accordingly, so as to achieve the purpose of removal. The gas escaping from the molten steel depends on the negative pressure of the vacuum chamber. With the gas escaping, the molten steel will boil and play the role of gas agitation. The reaction with gas phase can be controlled by adding aluminum or adjusting the pressure of the vacuum chamber. The degassing products gases, which are removed by entering CO and Ar blowing bubbles. Ar is blown into the molten steel through the porous permeable brick at the bottom of the ladle, and Ar forms small bubbles in the rising process, which is a vacuum chamber for H and N, so it diffuses and takes away. In addition, Ar blowing is beneficial to remove inclusions, uniform composition and temperature, and avoid secondary oxidation of molten steel. The pressure in the vacuum chamber is $666 \sim 2.66 \times 10^4$ Pa (5~200mmHg), and the treatment time is about 12~15min (determined by the temperature of the liquid steel). The smelting functions of VD Refining include:

VD 精炼主要与转炉、电弧炉和 LF 精炼配合,其基本原理基于 H、N 在钢液中溶解服从平方根定律,当 VD 抽真空时,真空室内压力降低,使 [H]、[N] 随之降低,达到去除目的。气体从钢液中逸出完全靠真空室的负压作用,伴随气体逸出,钢液产生沸腾,起到气体搅拌的作用,其间可用加铝或调节真空室压力的方法控制带气相的反应。脱气产物为气体,进入 CO 和吹氩气泡内被去除。通过钢包底部的多孔透气砖将 Ar 吹入钢液,Ar 在上升中形成小气泡,对 H、N 而言为真空室,因此向其中扩散并带走。此外,吹氩有利于去除夹杂,均匀成分和温度,避免钢液二次氧化。真空室内的压力为 $666 \sim 2.66 \times 10^4$ Pa($5 \sim 200$mmHg),处理时间大约 12~15min(决定于钢液温度)。VD 精炼能够实现的冶炼功能包括:

(1) Effective degassing, reducing [H] and [N].

(1) 有效脱气,减少 [H] 和 [N]。

(2) Deoxidation, C+[O]===CO.

(2) 脱氧,通过 C+[O]===CO 去除 [O]。

(3) Remove [S] by alkaline top slag.

(3) 通过碱性顶渣去 [S]。

(4) Chemical composition and temperature are controlled by alloy fine tuning and Ar blowing.

(4) 通过合金微调及吹 Ar 控制化学成分和温度。

(5) By blowing Ar, the inclusion will not gather and float up.

(5) 通过吹氩使夹杂无聚集上浮。

The main process flow is as follows: Ladle lifting into the tank→Start argon blowing→Temperature measurement sampling→Cover the vacuum tank→Start the vacuum pump→Adjust the vacuum degree and argon blowing intensity→Maintain the vacuum→Break the vacuum with nitrogen→Remove the tank cover→Temperature measurement sampling→Stop argon blowing and ladle lifting out of the station.

主要工艺流程为:吊包入罐→启动吹氩→测温取样→盖真空罐盖→开启真空泵→调节真空度和吹氩强度→保持真空→氮气破真空→移走罐盖→测温取样→停吹氩、吊包出站。

The sequence of vacuum treatment is exhaust→continuous compression→reaching the design vacuum degree (after about 6min)→keeping up with the VD treatment requirements→back pumping destroys the vacuum.

真空处理顺序为:排气→连续压缩→达设计真空度(约 6min 后)→保持达到 VD 处理要求→反抽气破坏真空。

When the ladle arrives at the working position, it will blow Ar, but not break the slag surface, so as to prevent the block of the permeable brick; When adding alloy, it will blow Ar with large flow, and blow the slag surface, so that the alloy can directly enter the molten steel and improve the recovery rate; When high vacuum treatment, it will blow Ar with small flow to prevent splashing; When feeding wire, it will blow Ar with small flow to prevent the increase of [N] and secondary oxidation.

当钢包到工位时,立即吹氩,但不吹破渣面,防止透气砖堵塞;当加合金时,采用大流量吹氩,吹开渣面,使合金直接进入钢液,提高收得率;当高真空处理时,小流量吹氩,防喷溅;当喂丝时,小流量吹氩,防止增加 [N] 和二次氧化。

Taking the production of GCr15 bearing steel as an example, the most important performance index of bearing steel is fatigue life. The important index affecting the life of bearing steel is the oxygen content in the steel, and it is better to control the [O] in the steel at 10ppm. The best level in the world [O] = 3~5ppm, about 10ppm in China. In addition, the grade of nonmetallic inclusions and carbides in steel was controlled by VD Refining.

以生产 GCr15 轴承钢冶炼为例，轴承钢最重要的性能指标是疲劳寿命。影响轴承钢寿命的重要指标是钢中氧含量，钢中 [O] 控制在 10ppm 为好。世界最好水平 [O] = 3~5ppm，国内 10ppm 左右。另外通过 VD 精炼控制钢中非金属夹杂物和碳化物级别。

The conditions of molten steel before refining (content of main elements) are: $w[C]$ = 0.95%~1.05%, $w[Mn]$ = 0.9%~1.20%, $w[Si]$ = 0.40%~0.65%, $w[Cr]$ = 1.30%~1.65%, $w[S]$< 0.02%, and $w[P]$<0.02%.

精炼前钢水条件（主要元素的含量）为：$w[C]$ = 0.95%~1.05%，$w[Mn]$ = 0.9%~1.20%，$w[Si]$ = 0.40%~0.65%，$w[Cr]$ = 1.30%~1.65%，$w[S]$<0.02%，$w[P]$<0.02%。

Smelting process is: UHP+LF+VD (or RH)+CC：

冶炼工艺是：UHP+LF+VD（或 RH）+CC：

The main process flow is as follows：

主要工艺流程如下：

(1) After LF tapping, 2/3 of slags (slag pouring) shall be removed, the thickness of slag layer shall be kept 40~70mm, and the time of slagging shall be less than 3min.

(1) LF 出钢后，扒渣（倒渣）2/3，渣层厚度应保持 40~70mm，扒渣时间小于 3min。

(2) After slagging, LF ladle is put into VD processing station, argon is connected, flow rate is adjusted to 50~80NL/min, temperature measurement and sampling are conducted at the same time, silica is added to 2kg/mm, and basicity of slag is adjusted to R = 1.2~1.5.

(2) 扒渣完毕后，LF 钢包入 VD 处理工位，接通氩气，调节流量 50~80NL/min，同时测温、取样，加入硅石 2kg/mm，调整炉渣碱度 R = 1.2~1.5。

(3) After temperature measurement and sampling, VD shall be sealed and vacuumized.

(3) 测温、取样后，VD 加盖密封，抽真空。

(4) During the start-up of vacuum pump, adjust the argon flow to maintain 30~40NL/min.

(4) 真空泵启动期间，调整氩气流量保持 30~40NL/min。

(5) Vacuum holding time: After vacuum startup, when the working pressure reaches 67 Pa, the holding time is no less than 15 min.

(5) 真空保持时间：真空启动后，当工作压力达到 67 Pa 时，保持时间不小于 15min。

(6) During the vacuum maintaining period, adjust the argon flow rate to about 70NL/min, and observe the boiling condition of the molten steel through the observation hole, adjusting in time, and keeping the uniform boiling.

(6) 真空保持期间调整氩气流量 70NL/min 左右，并通过观察孔观察钢水沸腾情况，及时调整，保持均匀沸腾。

(7) After the final deoxidation, the vacuum, cover opening and temperature measurement shall be removed, soft blowing shall be conducted for 15~25min, argon flow rate shall be about 70~100NL/min, and it is appropriate to control the micro motion of slag surface.

（7）终脱氧后，解除真空、开盖、测温，软吹 15~25min，氩气流量 70~100NL/min 左右，控制渣面微动为宜。

（8）After soft blowing, temperature measurement and sampling shall be carried out. Steel tapping shall be carried out with heat preservation agent, and the tapping temperature shall be 1530~1540℃.

（8）软吹结束后，测温、取样，加保温剂出钢，出钢温度 1530~1540℃。

2.8.2　Device of VD Refining

2.8.2　VD 精炼设备

VD the main refining equipment consists of vacuum system, vacuum tank system, vacuum tank cover car and feeding system, as shown in Figure 2-30. VD refining is mainly suitable for producing various alloy structural steel, high-quality carbon steel and low-alloy high-strength steel.

VD 精炼主要设备由真空系统、真空罐系统、真空罐盖车及加料系统组成，如图 2-30 所示。VD 精炼主要适于生产各种合金结构钢、优质碳钢和低合金高强度钢。

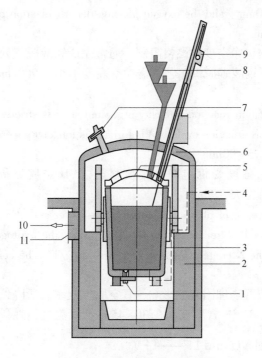

Figure 2-30　VD refining equipment
图 2-30　VD 精炼设备

1—Ladle argon blowing; 2—Vacuum chamber (tank); 3—Ladle; 4—Ladle support trunnion; 5—Charging cover; 6—Vacuum tank cover; 7—Vacuum pumping equipment; 8—Charging equipment; 9—Automatic temperature measurement and sampling device; 10, 11—Air breaking device

1—钢包吹氩；2—真空室（罐）；3—钢包；4—钢包支撑耳轴；5—加料盖；6—真空罐盖；7—抽真空设备；8—加料设备；9—自动测温取样装置；10，11—破空装置

2.8.3 VD Refining Limitations

2.8.3 VD 精炼局限性

VD treatment engineering, the oxygen content in steel can be reduced from 100 ppm to 20 ppm; The sulfur content (by mass) can be reduced from 0.01% to below 0.0015%, and the average desulfurization rate can reach 84%. However, due to the influence of static pressure of molten steel, the gas body at the bottom of ladle is not easy to escape. The effect is not significant, especially for the ladle with large tonnage. In addition, there are certain requirements for ladle equipment:

VD 精炼真空条件下实现钢—渣反应,有利于脱硫和脱氧。VD 处理工程中,钢中氧可从 100ppm 降低到 20ppm;硫含量(质量分数)可从 0.01% 降低到 0.0015% 以下,平均脱硫率可达 84%。但因受钢液静压力的影响,包底层的气体不易逸出,效果不大显著,特别是吨位较大的钢包。另外,对钢包设备的要求包括:

(1) The refining strength is strictly limited by the clearance of the ladle. Generally, the degassing treatment of the molten steel is completed, and the clearance height of the ladle is required to no less than 600mm; the carbon deoxidization process of the molten steel requires the clearance height is no less than 900mm; and the oxygen blowing decarbonization process requires the clearance height of 1.0~1.2m.

(1) 精炼强度受钢包净空度的严格限制,通常完成钢液脱气处理,要求钢包的净空高度不小于 600mm;进行钢液碳脱氧工艺,需净空高度不小于 900mm;吹氧脱碳,要求净空高度 = 1.0~1.2m。

(2) The decarburization reaction intensity of molten steel is strictly affected by the clearance height of ladle. The decarburization cycle of VD furnace is longer (generally 40~45min), and the whole treatment cycle is 75~90min.

(2) 钢水脱碳反应强度受钢包净空高度显著,VD 炉脱碳周期较长(一般为 40~45min),整个处理周期为 75~90min。

(3) Under the influence of physical factors such as cover and slag, the efficiency of vacuum degassing has been reduced. After degassing for 20~25min, the hydrogen content of molten steel can reach 2mg/L, and the N content fluctuates in 30~45mg/L. The consumption of argon is also higher than that of RH process.

(3) 受包盖、炉渣等物理因素的影响,真空脱气效率有所降低,脱气处理 20~25min,钢水氢含量可达到 2mg/L,N 含量波动在 30~45mg/L。氩气消耗量也高于 RH 工艺。

2.8.4 Combination of VD and LF——LFV

2.8.4 VD 与 LF 结合——LFV 精炼

In the past, the LF method mainly cooperated with the electric arc furnace. For the medium-sized converter workshop using continuous casting to produce common steel, it can only use the two functions of electric arc heating and bottom blowing argon stirring. Since 1990s, people are more and more interested in assembling LFV refining furnace in converter workshop. Using LFV refining method in steelmaking and continuous casting production line can reduce tapping temperature and iron oxide content in slag, and improve lining life, steel purity and casting rate of continuous casting. The special steel can be produced by oxygen converter with LFV instead of electric furnace.

过去LF法主要配合电弧炉，对于采用连铸生产普通钢的中型转炉车间，也可以只采用电弧加热和底吹氩搅拌两个功能。90年代以后，人们越来越喜欢在转炉车间装配LFV精炼炉。在炼钢与连铸生产线上采用LFV精炼法，可使出钢温度和炉渣中氧化铁含量降低，又可提高炉衬寿命，钢的纯净度，以及连铸的浇成率。可用氧气转炉配LFV法取代电炉法生产特殊钢。

The LVD method (Ladle Degassing) is a method of blowing argon into the ladle placed in a vacuum vessel. Because of its simple structure and easy maintenance, it is widely used in small-scale electric furnace plant for special steel refining. LF (V), Ladle Furnace (Vacuum), is the abbreviation of ladle furnace. The one without vacuum station is called LF method, and the one with vacuum station is called LFV method. LFV method is an improved equipment based on ASEA-SKF method and VAD method. These three methods are collectively called ladle refining furnace. The LVD method is to transfer the original reduction refining of EAF to the ladle for operation. The method of submerged arc barrel furnace is to heat the molten steel in the slag layer above the steel level, stir it with argon, and make slag in reducing atmosphere for refining.

LVD法（钢包脱气法）是一种向放在真空容器中的钢包里的钢水吹氩的方法。其结构简单，便于维护，广泛应用于小规模电炉厂进行特殊钢精炼。LF（V）是钢包炉的缩写。无真空工位的叫LF法，带有真空工位的叫LFV法。LFV法是在ASEA-SKF法和VAD法等方法基础上改进的设备，这三种方法统称为钢包精炼炉。此法是把电弧炉的还原精炼原样移到钢水包中操作。将电弧埋入钢液面以上的熔渣层中加热钢液，吹氩搅拌，在还原气氛下造渣精炼，也称为埋弧桶炉法。

Because LF method does not use vacuum, argon blowing is only for stirring, so degassing capacity is small. In order to degas, the original equipment is equipped with a vacuum cover and a vacuum chamber feeding equipment. This ladle furnace with vacuum degassing system is still called ladle refining furnace in China, as shown in Figure 2-31.

由于LF法未采用真空，吹氩只是为了搅拌，所以脱气能力小。为了脱气，在原设备上配备真空盖，并配有真空室下加料设备。这种带有真空脱气系统的钢包炉，我国仍称为钢包精炼炉，如图2-31所示。

2.8.4.1 Refining Process of LFV

2.8.4.1 LFV的精炼工艺

LFV can be matched with electric arc or converter, which can be in the same span or different span with electric furnace or converter. According to the characteristics and quality requirements of steel grades, LFV refining processes can be divided into four categories：

LFV既可以与电弧匹配，也可与转炉配合；既可与电炉或转炉处于同一跨中，也可以处于异跨中。根据钢种的特性及其质量要求，LFV的精炼工艺可以分为四类：

(1) Basic refining process：Electric furnace melting→Dephosphorization→Slagging-off→

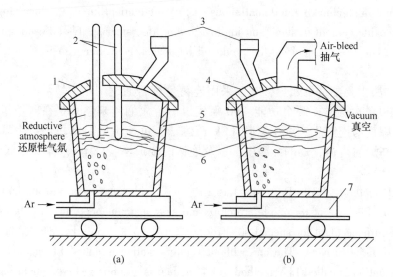

Figure 2-31 LFV method
图 2-31 LFV 法示意图
(a) Arc heating; (b) Vacuum treatment
(a) 电弧加热; (b) 真空处理

1—Heating cover; 2—Electrode; 3—Feeding tank; 4—Vacuum cover; 5—Ladle; 6—Alkaline reducing slag; 7—Ladle car
1—加热盖; 2—电极; 3—加料槽; 4—真空盖; 5—钢包; 6—碱性还原渣; 7—钢包车

Slagging forming→Alloying→Tapping→Refining ladle entering LFV ladle station→Argon blowing→Heating→Adjusting composition→Vacuum degassing composition→Fine tuning→Ladle casting.

This process is suitable for the production of pure steel, with refining time of 50~70min and power consumption of 30~40kW·h/t.

(1) 基本精炼工艺为: 电炉熔化→脱磷→扒渣→造渣→合金化→出钢→精炼钢包进入 LFV 座包工位→吹氩→加热→调整成分→真空脱气→成分微调→吊包浇注。

这种工艺适用于纯净钢的生产, 其精炼时间 50~70min, 电耗 30~40kW·h/t。

(2) Special refining process: Electric furnace melting and composition analysis→Dephosphorization→Slagging→Tapping→Ladle alloying entering LFV heating station→Slagging→Heating→Argon blowing vacuum→Degassing vacuum→Alloying composition and temperature→Fine tuning→Ladle pouring.

The refining time is 70~90min, and the power consumption is 40~50kW·h/t.

(2) 特殊精炼工艺为: 电炉熔化, 成分分析→去磷→扒渣→出钢→钢包合金化→进入 LFV 加热工位→造渣, 加热, 吹氩→真空脱气→真空合金化→成分和温度微调→吊包浇注。

这种工艺适用于超纯净钢生产, 其精炼时间 70~90min, 电耗 40~50kW·h/t。

(3) General refining process includes: Electric furnace melting and composition analysis→Dephosphorization→Slag raking→Slagging and alloying→Tapping→Ladle into LF (V) ladle Station→Argon blowing and heating→Composition and temperature adjustment→Ladle casting.

This process is suitable for low alloy steel without vacuum. The refining time is 30min and the power consumption is 30~40kW·h/t.

(3) 普通精炼工艺为: 电炉熔化, 成分分析→去磷→扒渣→造渣, 合金化→出钢→钢包进 LF（V）座包工位→吹氩，加热→成分和温度调整→吊包浇注。

这种工艺适用于一般要求低合金钢，无真空。其精炼时间在 30min，电耗 30~40kW·h/t。

(4) Vacuum oxygen blowing decarburization process: Electric furnace melting and composition analysis→Oxygen blowing and decarburization→Initial reduction→Composition adjustment→Tapping→Ladle slag removal→Slag addition→Heating and argon blowing→Vacuum oxygen blowing and decarburization→Vacuum alloying→Vacuum degassing→Composition and temperature fine adjustment→Ladle pouring.

This process is suitable for the production of low carbon and ultra-low carbon stainless steel. The refining time is 120~150min and the power consumption is 20~30kW·h/t.

(4) 真空吹氧脱碳工艺为：电炉熔化，成分分析→吹氧，脱碳→初还原→调整成分→出钢→钢包除渣→加渣料→加热，吹氩→真空吹氧，脱碳→真空合金化→真空脱气→成分和温度微调→吊包浇注。

这种工艺适用于生产低碳和超低碳不锈钢，其精炼时间 120~150min，电耗 20~30kW·h/t。

2.8.4.2　Refining Equipment of LFV

2.8.4.2　LFV 的精炼设备

LFV is usually composed of the seat bag station, heating station and vacuum station, which can be arranged in either the Seat bag—Heating—Vacuum type or the Seat bag—Vacuum—Heating type.

LFV 通常由座包工位，加热工位，真空工位组成，既可以座包—加热—真空型式布置，也可以座包—真空—加热形式布置。

The vacuum chamber of LFV has two types of structure: The vacuum cover and refining ladle are directly sealed with heat-resistant rubber sealing ring, that is, the barrel type sealing structure; The vacuum tank and the vacuum tank cover form a closed vacuum chamber, which is a tank type sealing structure. The former is suitable for medium and small LFV under the existing plant conditions. The advantages are: small floor area and flexible operation, but requiring high requirements for the shape and size of the ladle mouth. The latter is more suitable for refining low carbon and ultra-low carbon steel. Moreover, there is no special requirement for ladle, but the floor area and vacuum volume are relatively large.

LFV 的真空室有两种结构形式：真空包盖与精炼钢包直接用耐热橡胶密封圈密封，即桶式密封结构；真空罐与真空罐盖组成一个密闭的真空室，即罐式密封结构。前者适合于现有厂房条件的中、小型 LFV，其优点是：占地面积小，操作较灵活，但对精炼钢包的包口外形尺寸要求比较高。后者比较适合于低碳和超低碳钢的精炼，而且对钢包没有特殊要求，但占地面积和真空体积相应都比较大。

The vacuum pump used in LFV furnace is the same as other ladle refining furnaces, most of which are steam jet pumps. Compared with mechanical pumps, jet pumps are more suitable for metallurgical process, because they do not need to worry about the exhaust temperature, extract small slag particles and metal dust in the gas, and also have a huge exhaust capacity that mechanical pumps cannot match. But the steam pump needs a lot of cooling water and steam. The pumping capacity of steam jet pump is generally selected according to the factors of liquid steel treatment, steel treatment, refining process, treatment time, vacuum volume, etc. For example, for LFV (30~50t), the capacity of the steam jet pump with barrel vacuum structure is generally 150kg/h, while the capacity of the steam jet pump with tank vacuum structure is generally 250kg/h. The limiting vacuum degree of LFV is generally 67~270Pa.

LFV 炉所采用的真空泵同其他的钢包精炼炉一样,多数为蒸汽喷射泵。与机械泵相比,喷射泵更适用于冶金过程,因为它不必顾虑排气温度,抽出气体中的微小渣粒及金属尘埃等,而且还具有机械泵无法比拟的巨大排气能力。但是蒸汽泵需要大量的冷却水和蒸汽。蒸汽喷射泵抽气能力一般根据处理钢液量、处理钢种、精炼工艺、处理时间、真空体积等因素来选择。例如对于 LFV(30~50t),处理一般纯净度要求的钢种,采用桶式真空结构的蒸汽喷射泵的能力一般为150kg/h,而采用罐式真空结构的蒸汽喷射泵的能力一般为250kg/h。LFV 的极限真空度一般为67~270Pa。

2.8.4.3 Smelting Function

2.8.4.3 冶炼功能

LFV method has the same functions as VAD, such as reducing atmosphere in furnace, bottom blowing argon stirring, submerged arc heating of graphite electrode under atmospheric pressure, refining of high basicity synthetic slag, fine adjustment of alloy composition and vacuum degassing. The difference is that the vacuum and heating are respectively covered with two covers, which are heated under atmospheric pressure, refined with synthetic slag, stirred with argon blowing, and then vacuumed for degassing. VAD heating and vacuum with a cover and vacuum arc heating, can be added in vacuum alloy fine-tuning alloy composition.

LFV 法的功能与 VAD 相同,炉内还原性气氛,底吹氩气搅拌,大气压下石墨电极埋弧加热,高碱度合成渣精炼,微调合金成分,真空脱气。不同之处在于真空和加热分别采用两个包盖,大气压下加热,加合成渣精炼,吹氩搅拌,然后抽真空脱气。而 VAD 的加热和真空同用一个包盖,真空下电弧加热,可以在真空下加合金微调合金成分。

2.8.4.4 Refining Effect

2.8.4.4 精炼效果

LFV refining can achieve refining effects including degassing, deoxidization, decarburization, desulfurization, inclusion removal, heating molten steel, fine-tuning composition, etc. If equipped with an oxygen lance, it can also decarburize by vacuum oxygen blowing and smelt stainless steel. Using vacuum argon blowing can make $w[H] \leq 2.68 \times 10^{-4}\%$, $w[N] \leq 37 \times 10^{-4}\%$, and $w[O] \leq 10 \times 10^{-4}\%$. The [S] of industrial pure iron (by mass) can be reduced from 0.060%

to less than 0.015% by using common process. The [S] in bearing steel can be reduced from 0.030% to less than 0.003% by special refining process. The special vacuum refining and argon blowing stirring system can not only make the [S]+[H]+[N]+[O] in bearing steel less than 70×10^{-4}%. And the oxide content of steel is less than 0.003% (by mass), and the sulfide content is less than 0.0246% (by mass). The temperature range of molten steel can be controlled within 2.5℃.

LFV 精炼可以实现包括脱气、脱氧、脱碳、脱硫、去夹杂、加热钢液、微调成分等精炼效果。如果配一支吹氧枪还可以真空吹氧脱碳，冶炼不锈钢。采用真空吹氩可使轴承钢的 $w[H]\leq2.68\times10^{-4}$%，$w[N]\leq37\times10^{-4}$%，$w[O]\leq10\times10^{-4}$%。采用普通工艺可使工业纯铁的（质量分数）[S] 从 0.060% 下降到 0.015% 以下。采用特殊精炼工艺可使轴承钢中的 [S] 从 0.030% 下降到 0.003% 以下。采取特殊的真空精炼和吹氩搅拌制度不仅可使轴承钢中的 [S]+[H]+[N]+[O]$\leq70\times10^{-4}$%。而且钢中氧化物含量（质量分数）达到 0.003% 以下，硫化物含量（质量分数）达到 0.0246% 以下。钢液温度范围可控制在 2.5℃ 内。

Task 2.9　VAD Refining and VOD Refining
任务 2.9　VAD 精炼与 VOD 精炼

VAD (Vacuum Arc Degassing) refining uses ladle vacuum arc heating degassing method, which is called arc heating degassing method or vacuum arc degassing method. It uses argon to stir molten steel and adds arc heating device on the cover of vacuum chamber. The difference between this method and the ladle refining furnace method (ASEA-SKF method) is that it stirs the liquid steel with argon and heats it with electric arc under vacuum. VOD (Vacuum Oxygen Decarburization) was developed by witten in 1965. The composition of VOD equipment and VD equipment is basically the same, and the main difference is that VOD method adds oxygen gun, its lifting system and oxygen supply system. Both methods are innovative technologies developed on the basis of VD refining.

VAD 精炼是利用钢包真空电弧加热脱气，也可称为电弧加热钢包脱气法或真空电弧钢包脱气法。其是用氩气搅拌钢液，并在真空室的盖子上增设电弧加热装置。这种方法与钢包精炼炉法（ASEA-SKF 法）的不同之处在于它用氩气搅拌钢液，并在真空下进行电弧加热。VOD 是 1965 年由德国维腾公司开发出的技术，其与 VD 设备的构成基本相同，主要的区别在于 VOD 法增加了氧枪及其升降系统、供氧系统。这两种方法都是在 VD 精炼基础上开发的创新技术。

2.9.1　VAD Process Characteristics
2.9.1　VAD 工艺特点

The process features are as follows:

VAD 精炼工艺特点如下：

(1) Since the arc heating is carried out in vacuum, and the vacuum heating is about $(2.4 \sim 2.6) \times 10^4$ Pa ($180 \sim 120$ mmHg), and good degassing effect can be obtained in the heating process.

(1) 由于电弧加热是在真空下进行的，真空加热约在 $(2.4 \sim 2.6) \times 10^4$ Pa ($180 \sim 120$ mmHg) 进行，故在加热过程中可以获得良好的脱气效果。

(2) The pouring temperature can be adjusted accurately, the ladle lining has sufficient heat storage, and the temperature drop is stable during pouring.

(2) 能够准确地调整浇注温度，且钢包内衬充分蓄热，浇注时温降稳定。

(3) The composition of the molten steel is stable because of the sufficient agitation in the refining process.

(3) 由于精炼过程中搅拌充分，所以钢液成分稳定。

(4) A large number of alloys can be added and a wide range of carbon steel and alloy steel can be smelted.

(4) 可加入大量的合金，能冶炼范围很广的碳素钢与合金钢。

(5) Slag agent and other slag materials can be added for desulfurization and decarburization. If an oxygen gun is installed on the vacuum cover, the vacuum oxygen decarburization process can also be used to smelt ultra-low carbon stainless steel.

(5) 可以加入造渣剂和其他造渣材料进行脱硫和脱碳。如果在真空盖上装设氧枪，还可采用真空吹氧脱碳工艺，冶炼超低碳不锈钢。

The differences between VAD and ASEA-SKF method are as follows:

VAD 与 ASEA-SKF 法的不同之处在于：

(1) Stirring with argon, rather than electromagnetic is more intense. In order to maintain the boiling time of vacuum degassing, there is a large free space in the upper part of the ladle.

(1) 用氩气搅拌，而不是用电磁搅拌，搅拌更激烈，为了保持真空脱气的沸腾时间，钢包上部留有较大的自由空间。

(2) The vacuum cover is equipped with a double bell funnel which can add alloy and flux under vacuum, so as not to damage the vacuum degree during continuous feeding.

(2) 真空盖上设有可在真空下添加合金及熔剂的双钟式漏斗，以便连续加料时不破坏真空度。

(3) The vacuum cover is also equipped with arc heating device. The key part of the equipment is the sealing technology between the graphite electrode moving up and down and the vacuum cover. The telescope type double sleeve system is adopted, and the sleeve is made of antimagnetic material and cooled with water.

(3) 真空盖上还增设了电弧加热装置，设备的关键部位是上下移动的石墨电极与真空盖之间的密封技术，采用了望远镜式的双套筒系统，套筒用抗磁材料制成，并用水冷却。

2.9.2 The Main VAD Equipment

2.9.2 VAD 主要设备

The main equipment of VAD refining includes vacuum chamber, argon blowing equipment, heating system, feeding system, etc., as shown in Figure 2-32.

VAD 精炼的主要设备包括真空室、吹氩设备、加热系统、加料系统等。设备示意图如图 2-32 所示。

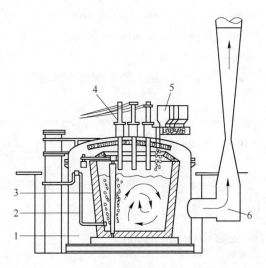

Figure 2-32 VAD method equipment diagram
图 2-32 VAD 法设备示意图
1—Vacuum chamber; 2—Bottom argon blowing system; 3—Ladle; 4—Arc heating system;
5—Alloy feeding system; 6—Vacuum pumping device
1—真空室;2—底吹氩系统;3—钢包;4—电弧加热系统;5—合金加料系统;6—抽真空装置

VAD refining can carry out ladle degassing and ladle refining, degassing treatment is about 15min. At the same time of vacuumizing, the bottom of the ladle is blown with argon to stir the molten steel. A VAD furnace with a capacity of 50t needs 35kW·h/t to heat up the liquid steel at 50℃, and the vacuum refining time is generally not less than 30min. After arc heating, it can compensate 100~200℃. In addition, when the VAD refining system is generally equipped with a water-cooled oxygen gun, the vacuum oxygen blowing decarburization process can be used, which is of great significance for refining stainless steel. The yield of chromium is about 98%, and its quality is the same as VOD.

VAD 精炼可进行钢包脱气和钢包精炼,脱气处理约 15min。抽真空的同时,包底吹氩搅拌钢液。一个容量为 50t 的 VAD 炉,使钢液升温 50℃ 需耗电 35kW·h/t,真空精炼处理时间一般不小于 30min。电弧加热后可补偿 100~200℃。另外,在 VAD 精炼系统一般设有水冷氧枪时可采用真空吹氧脱碳工艺,这对精炼不锈钢具有重要意义。铬的收得率约为 98%,其质量与 VOD 法相同。

2.9.3 Refining Effect

2.9.3 精炼效果

VAD refining method has many metallurgical functions, such as vacuum degassing, vacuum arc heating, argon blowing and stirring, slag forming and refining, which can realize:

VAD 精炼法具有真空脱气、真空下电弧加热、吹氩搅拌、造渣精炼多种冶金功能，能够实现：

(1) The average hydrogen content of molten steel after treatment is 1.3ppm, and the dehydrogenation rate is 65%. The dehydrogenation effect is remarkable. The hydrogen content of 92% molten steel in ingot mold is less than 2ppm.

（1）钢液处理后氢含量平均为 1.3ppm，脱氢率为 65%，脱氢效果显著。锭模内 92% 钢液氢含量小于 2ppm。

(2) The average oxygen content in the molten steel is 24ppm, the deoxidization rate is 54%, the oxygen recovery after treatment is about 5ppm, and the total oxygen content in 90% of the molten steel is below 30ppm.

（2）钢液中的氧平均含量为 24ppm，脱氧率为 54%，处理后氧回升约 5ppm，90% 的钢液中总含氧在 30ppm 以下。

2.9.4 VOD Refining Process Characteristics

2.9.4 VOD 精炼工艺特点

VOD refining can be smelted in vacuum. The purity of steel is high and the content of carbon and nitrogen is low. Various special steels are treated by vacuum refining or degassing. The bottom of the ladle is agitated by argon blowing. The degassing effect is better and the metallurgical reaction kinetics is more favorable. VOD method is to blow oxygen from furnace top to molten steel in vacuum chamber, while stirring molten steel by blowing argon from ladle bottom. When the refining meets the decarburization requirements, stop blowing oxygen, then increase the vacuum degree for deoxidation, and finally add Fe-Si for deoxidation. It can add alloy, sample and measure temperature under vacuum. Because of the strong carbon oxygen reaction, the height of the free space above the ladle is required to be 1.0~1.2m. The refining process features include:

VOD 精炼可在真空条件下冶炼，钢的纯净度高，碳和氮的含量低。对各种特殊钢进行真空精炼或真空脱气处理，钢包底部设有吹氩搅拌，其脱气去夹杂的效果较好，冶金反应动力学较为有利。VOD 法是在真空室内由炉顶向钢液吹氧，同时由钢包底部吹氩搅拌钢水，当精炼达到脱碳要求时，停止吹氧，然后提高真空度进行脱氧，最后加 Fe-Si 脱氧。其可以在真空下加合金，取样和测温。因为强烈的碳氧反应，要求钢包上部的自由空间的高度为 1.0~1.2m，精炼工艺特点包括：

(1) The content of C is 0.4%~0.5% (by mass), and the liquid steel temperature is 1600~1650℃ during tapping. Pouring the molten steel into the ladle, the slag shall not flow into the la-

dle as much as possible.

(1)出钢时碳含量(质量分数)为 0.4%~0.5%,钢液温度 1600~1650℃。将钢水倾入钢包,尽量避免钢渣流入包内。

(2) The ladle with molten steel is hoisted into the vacuum tank to blow argon and stir while vacuumizing. When the pressure is 6700Pa (50mmHg), oxygen blowing refining begins. At this time, the amount of slag on the surface of the molten pool should be less.

(2)装有钢水的钢包吊到真空罐内吹氩搅拌边抽真空,压力 6700Pa(50mmHg)时开始吹氧精炼。此时,熔池表面上的渣量少些为宜。

(3) The vacuum degree is about 1kPa. The exothermic of oxidation reaction makes the temperature of molten steel increase slightly.

(3)真空度达 1kPa 左右。氧化反应放热,使钢液温度略有升高。

(4) Deoxidize in vacuum or atmosphere, lift the ladle out of the vacuum tank and pour after adjusting the composition and temperature.

(4)在真空下或大气中进行脱氧。经调整成分和温度后,从真空罐内将钢包吊出,进行浇注。

(5) Deoxidization of refined stainless steel: Excess carbon is deoxidized and inclusions are removed under vacuum of less than 130Pa. The purity of stainless steel deoxidized by aluminum and titanium is also significantly improved by stirring in argon under vacuum.

(5)精炼不锈钢的脱氧:过剩的碳在小于 130Pa 的真空下脱氧和去除夹杂物。在真空下用氩气搅拌,因此用铝、钛脱氧的不锈钢的纯洁度也显著提高。

(6) The equipment is complex, the smelting cost is high, the decarburization speed is slow, the primary furnace needs to carry out rough decarburization, and the production efficiency is low.

(6)设备复杂,冶炼费用高,脱碳速度慢,初炼炉需要进行粗脱碳,生产效率低。

2.9.5　VOD Refining Process

2.9.5　VOD 精炼过程

VOD process has the functions of decarburization, deoxidation, degassing, desulfurization and alloying. VOD Refining can continuously reduce the partial pressure of CO in the molten steel environment and achieve carbon removal and chromium retention. It is mainly used in the production of stainless steel or ultra-low carbon alloy steel. Take the smelting of ultra-low carbon stainless steel as an example:

VOD 法具有脱碳、脱氧、脱气、脱硫及合金化等功能。VOD 精炼可以不断降低钢水所处环境的 CO 的分压力,达到去碳保铬。其主要用于生产不锈钢或超低碳合金钢,以冶炼超低碳不锈钢为例:

(1) The content of C is between 0.2% and 0.5% (by mass) in primary furnace, and the content of P is not higher than 0.03%. The temperature of molten steel is 1630℃.

(1)初炼炉将 C 含量(质量分数)控制在 0.2%~0.5%之间,P 含量(质量分数)在 0.03%以下;钢液温度为 1630℃。

(2) After the slag removed from the primary smelting furnace, the VOD ladle is hoisted into the vacuum chamber. Then argon is blown at the bottom, and vacuum is started. At this time, the temperature is 1550~1580℃.

(2) 初炼炉除渣后，将 VOD 钢包吊入真空室，接底吹氩，开始抽真空。此时温度为 1550~1580℃。

(3) When the vacuum degree reaches 13~20kPa, oxygen blowing and decarbonization are started.

(3) 当真空度达到 13~20kPa 时，开始吹氧脱碳。

(4) At the same time of reducing the carbon content, improve the vacuum degree and protect chromium from oxidation.

(4) 碳含量降低的同时，提高真空度，保铬不氧化。

(5) When the carbon qualified, stop blowing oxygen, increase the vacuum to below 100Pa, increase the agitation and further decarburization, and the temperature of liquid steel reaches 1670~1750℃.

(5) 当碳合格时，停止吹氧，加大真空到 100Pa 以下，并加大搅拌，进一步脱碳，钢液温度达到 1670~1750℃。

(6) Add alloy, fine tune composition, add aluminum and blow argon and stir for several minutes, and then break vacuum casting.

(6) 加合金，微调成分，加铝吹氩搅拌几分钟后，破真空浇铸。

2.9.6 VOD Refining Equipment

2.9.6 VOD 精炼设备

The main equipment of VOD Refining includes: ladle (clearance height 1000~1200mm; good refractory material); vacuum tank (Oxygen gun, temperature measuring sampling device and feeding device shall be installed on the cover of vacuum tank; Water-cooled slag baffle plate shall be set inside); vacuum system (water ring pump and steam jet pump); oxygen gun (consumable steel pipe; laval spray gun), etc., as shown in Figure 2-33.

VOD 精炼的主要设备包括：钢包（净空高 1000~1200mm，耐火材料好），真空罐（真空罐的盖子上要安装氧枪、测温取样装置、加料装置、内部要设置水冷挡渣盘），真空系统（水环泵和蒸汽喷射泵），氧枪（自耗钢管；拉瓦尔喷枪）等。设备示意图如图 2-33 所示。

Compared with other ladle degassing methods, VOD Refining Equipment is similar except for two differences. One is that the oxygen gun for oxygen blowing refining can be lifted and lowered freely through the cover of vacuum chamber. The other is that a large amount of CO gas is produced by oxygen blowing and decarburization in vacuum chamber. So the ability of air extraction should be enhanced. VOD method has outstanding advantages for refining ultra-low carbon steel, and it is also very effective for vacuum refining or vacuum degassing of various special steels. Because the bottom of the ladle is equipped with an argon stirring device, the degassing effect is bet-

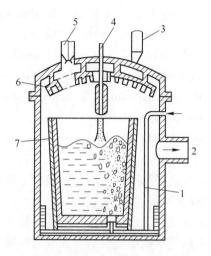

Figure 2-33　VOD refining equipment
图 2-33　VOD 精炼设备示意图

1—Argon blowing device; 2—Degassing vacuum chamber; 3—Charging device; 4—Oxygen blowing device;
5—Sampling and temperature measuring device; 6—Protective cover; 7—Ladle
1—吹氩装置；2—脱气真空室；3—加料装置；4—吹氧装置；5—取样和测温装置；6—保护盖；7—钢包

ter and the metallurgical reaction kinetic conditions are better. If the treatment time is short and the ladle is preheated properly, the temperature drop of the molten steel is not large and the effect is good, although the molten steel is not heated up during the treatment. However, the ladle for refining bears four functions: vacuum oxygen blowing, decarburization, refining and pouring. In order to prevent splashing during oxygen blowing and decarburization, the height and diameter of ladle is larger than that of general ladle.

VOD 精炼设备与其他钢包脱气法比较，除有两点不同外其他均相似。一是吹氧精炼用的氧枪可通过真空室盖子自由升降；另一点是由于真空室进行吹氧脱碳产生大量 CO 气体，故要增强抽气能力。VOD 法对于精炼超低碳钢种具有突出的优越性，同时对各种特殊钢进行真空精炼或真空脱气处理也很有效。该法由于包底设有吹氩搅拌装置，其脱气去除夹杂的效果比较好，冶金反应动力学条件也较好。如果处理时间短，钢包采用适当的预热措施，钢水在处理过程中虽未进行补充加热，钢水的温降不大，效果较好。精炼用的钢包承担真空吹氧、脱碳、精炼和浇注四个功能。为了防止吹氧脱碳时产生喷溅，钢包的高度直径比要比一般钢包大些。

Task 2.10　DH Refining

任务 2.10　DH 精炼

The development trend of the new generation of steel materials is super clean, high uniformity and microstructure control. The degassing refining of steel flow can meet the requirements of cleanliness of all kinds of high quality steel. The degassing method is to inject the molten steel stream into the vacuum chamber. Due to the sharp drop of pressure, the stream expands suddenly and spreads to a certain angle to drop down, which greatly increases the degassing surface area and is conducive to the escape of gas. The study on the effect of drop deoxidation and decarburization on the final deoxidized molten steel under vacuum has proved that in the process of vacuum drop deoxidation, decarburization also occurs, and the content of non-metallic inclusions also decreases by about 50% ~ 70%. Steel flow degassing method includes vacuum lift degassing method DH and vacuum circulation degassing method RH. This task first introduces the vacuum lifting degassing method——DH refining.

新一代钢铁材料的发展趋势是超洁净、高均匀和微细组织结构控制。钢流脱气精炼可以满足各类高品质钢材洁净度的要求。钢流脱气法是将钢水流股注入真空室，由于压力急剧下降，使流股突然膨胀并散开成一定角度以滴状降落，使脱气表面积大大增加，有利于气体逸出。经真空下对未脱氧的钢水钢滴流脱氧和脱碳的作用研究证明，在真空滴流过程中，钢水脱氧的同时还有脱碳作用。另外，非金属夹杂物的含量（质量分数）也相应降低约50% ~ 70%。钢流脱气法包括真空提升脱气法DH和真空循环脱气法RH。本任务首先介绍真空提升脱气法——DH精炼。

2.10.1 The Equipment of DH Refining

2.10.1 DH 精炼设备

DH vacuum lift degassing method was first invented and used by Dortmund and Horder Metallurgical Joint Company in 1956, which is called DH method for short. As shown in Figure 2-34, it is composed of vacuum chamber (steel shell lined with refractory) and lifting mechanism, heating device (electrode heating device or gas injection, oil spray heating), alloy silo (sealed charging under vacuum), air extraction system, etc. It mainly decarbonizes, deoxidizes, denitrifies and other gases; decarbonizes in vacuum; stirs for homogenization of composition and temperature and separate non-metallic inclusions.

DH 真空提升脱气法是1956年德国多特蒙特和蒙特尔冶金联合公司首先发明使用的，简称DH法。如图2-34所示，DH精炼设备由真空室（钢壳内衬耐火材料）及提升机构，加热装置（电极加热装置或喷燃气、喷油加热），合金料仓（真空下密封加料），抽气系统等组成。主要进行脱碳、脱氧、脱氮等气体；真空脱碳；搅拌进行成分、温度的均匀化和分离非金属夹杂物等。

2.10.2 DH Refining Process

2.10.2 DH 精炼工艺

2.10.2.1 Principle of DH Refining Degassing

2.10.2.1 DH 精炼脱气工作原理

According to the principle of pressure balance, with the help of the relative movement

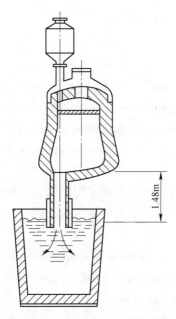

Figure 2-34　DH refining equipment
图 2-34　DH 精炼设备示意图

between the vacuum chamber and the ladle, the liquid steel is sucked into the vacuum chamber through the suction nozzle in batches for degassing treatment. During the treatment, the suction nozzle at the lower part of the vacuum chamber is inserted into the liquid steel. After the vacuum chamber is vacuumized, the pressure difference is formed inside and outside. The liquid steel rises to the height of the pressure difference in the vacuum chamber along the suction nozzle. If the indoor pressure is 13.3~66Pa, the liquid steel will be raised about 1.48m. As the liquid steel in the vacuum chamber boils to form droplets, the Gas—Liquid interface area is greatly increased, and the gas in the steel is removed due to the vacuum effect. When the ladle is lowered or the vacuum chamber is lifted, the degassed liquid steel returns to the ladle. When the ladle is lowered or the vacuum chamber is lowered, another batch of liquid steel enters the vacuum chamber for degassing, so that the liquid steel enters the vacuum chamber batch by batch until the end of treatment.

　　根据压力平衡原理，借助于真空室与钢包之间的相对运动，将钢液经吸嘴分批吸入真空室内进行脱气处理的。处理时将真空室下部的吸嘴插入钢液内，真空室抽成真空后其内外形成压力差，钢液沿吸嘴上升到真空室内的压差高度，如果室内压力为 13.3~66Pa，则提升钢液约 1.48m。由于真空作用室内的钢液沸腾形成液滴，大大增加气液相界面积，钢中的气体由于真空作用而被脱除。当钢包下降或真空室提升时脱气后的钢液重新返回到钢包内。当钢包下降或真空室下降时又有一批钢液进入真空室进行脱气，这样钢液一批一批地进入真空室直至处理结束为止。

2.10.2.2 DH Refining Operation Process

2.10.2.2 DH精炼操作工艺

Preparation stage: Before treatment, the type and quantity of alloy is determined to be added according to the steel type and tapping amount. And it is added into the silo in advance, the steel amount is selected to be inhaled each time according to the steel amount and ladle size, the lifting stroke and limit position are adjusted, the slag retaining cap is installed before the nozzle, and the vacuum chamber is heated.

准备阶段：处理前根据钢种和出钢量确定配加的合金种类和数量，并预先加入到料仓内，根据钢水量和钢包尺寸选定每次吸入的钢水量，调整好升降行程和极限位置，吸嘴前装好挡渣帽，将真空室加热。

Treatment stage: After the ladle containing liquid steel is sent to the treatment position, temperature measurement and sampling are carried out, and the suction nozzle is inserted into the liquid steel, and then the vacuum pump is started to pump air. When the pressure of the vacuum chamber drops to 1.333×10^4 Pa, the lifting mechanism starts to rise and fall automatically. And the liquid steel entering the vacuum chamber starts degassing reaction under the effect of low pressure, resulting in intense boiling and splashing. After degassing, the molten steel flows back to the ladle to produce violent mixing and even mixing. In this way, 30 times of lifting and lowering are carried out repeatedly. After 3 times of circulation, the vacuum degree is stabilized to the limit value, and the alloy is added, and then the lifting and lowering are carried out several times. After the alloy is fully mixed, the sample temperature is measured and sent to the pouring.

处理阶段：将盛有钢液的钢包送到处理位置后测温、取样并将吸嘴插入钢液内，然后启动真空泵抽气。当真空室压力降至 1.333×10^4 Pa 时，升降机构开始自动升降，进入真空室的钢液在低压作用下开始脱气反应，产生激烈的沸腾和喷溅。脱气后钢液回流到钢包内产生剧烈的搅拌和混匀。这样反复进行 30 多次左右的升降，全部钢液经 3 次循环，真空度稳定到极限值，然后加入合金，再升降几次待合金充分混匀后取样测温送去浇注。

The main process parameters are: liquid steel suction, lifting times, circulation factor, pause time, lifting speed, lifting stroke, etc. Reasonable determination of these parameters is of great significance to the design of the vacuum chamber and the determination of the process system to achieve satisfactory degassing effect.

主要工艺参数有：钢液吸入量，升降次数，循环因数，停顿时间，升降速度，提升行程等。合理确定这些参数对真空室的设计，确定工艺制度对达到满意的脱气效果具有重要意义。

2.10.2.3 Refining Effect of DH

2.10.2.3 DH精炼效果

After DH treatment, hydrogen, nitrogen, oxygen and non-metallic inclusions in the steel have been reduced, internal and external defects of the steel have been significantly reduced, and various properties have been improved. The refining effects are as follows:

DH 处理后钢中的氢、氮、氧及非金属夹杂物都有相当的减少，钢材内外部缺陷也明显减少，各种性能也得到了提高。精炼效果如下：

(1) The effect of dehydrogenation is good, which can be reduced from 2.5~6.5ppm to 1.0~2.5ppm. When the undeoxidized steel is treated, a large number of CO bubbles are produced from the bottom of the molten pool, which is conducive to the dehydrogenation reaction.

(1) 脱氢效果较好，可由处理前的 2.5~6.5ppm 降低到 1.0~2.5ppm。当处理未脱氧钢时，熔池底部产生大量 CO 气泡，有利于脱氢反应的进行。

(2) The oxygen can be reduced by 55%~90% and the non-metallic inclusions can be reduced by 40%~50% by treating the steel without pre deoxidization. When the alloy is added under vacuum, the yield can reach more than 95%, and the composition and temperature are uniform.

(2) 处理未经预脱氧钢液可使氧降低 55%~90%，还可以降低非金属夹杂物 40%~50%。合金在真空下加入，收得率高达 95% 以上，并且成分和温度均匀。

(3) The effect of denitrification is poor. When the nitrogen content in steel is low (30~40ppm), there is almost no change in short time treatment. When the nitrogen content is higher than 100ppm, the denitrification amount can reach 20~30ppm.

(3) 脱氮效果较差。当钢中氮含量低 (30~40ppm) 时，短时间处理几乎无变化；当氮含量高于 100ppm 时，脱氮量可达 20~30ppm。

(4) During vacuum treatment, the DH method can produce ultra-low carbon steel with $w[C] \leq 0.002\%$ with the reduction of carbon content due to carbon oxygen reaction.

(4) 真空处理时，由于碳氧反应降低了碳含量，因此 DH 法可生产超低碳钢 $w[C] \leq 0.002\%$。

In a word, after DH refining, a smaller vacuum chamber can be used to treat large tonnage of liquid steel. Graphite electrode, heavy oil and gas can also be used to bake and heat the vacuum chamber, so the temperature drop of liquid steel is smaller. Due to the intense boiling, it also has a large decarburization capacity, which can be used to produce low carbon steel containing $w[C] < 0.002\%$. Alloy can be added in the process of treatment, and the recovery rate of alloy elements is high. However, DH refining equipment is relatively complex, with high investment and operation costs, which is currently suitable for large capacity smelting equipment.

总之，经 DH 精炼可以使用较小的真空室处理大吨位的钢液，可以用石墨电极、重油和煤气对真空室进行烘烤和加热，因此钢液温降较小。由于激烈的沸腾还具有较大的脱碳能力，可以用来生产 $w[C] \leq 0.002\%$ 的低碳钢，处理过程中可以加合金，合金元素的收得率高。但 DH 精炼设备比较复杂，投资和操作费用都比较高，目前适用于大容量的冶炼设备。

Task 2.11　RH Refining
任务 2.11　RH 精炼

For steel grades (such as electrical silicon steel) requiring ultra-low carbon and ultra-low sulfur at the same time, steel grades (such as IF steel) requiring ultra-low carbon and ultra-low nitrogen at the same time, and steel grades (such as coated steel plate) requiring low carbon and low silicon at the same time, RH is the only best refining equipment. RH refining was jointly invented by Ruhrstahl A. G. and Heraeus A. G in 1957, so it is called RH method for short.

对于同时要求超低碳、超低硫的钢种（如电工硅钢）和同时要求超低碳、超低氮的钢种（如 IF 钢）以及同时要求低碳、低硅的钢种（如涂镀钢板），RH 是唯一最佳的精炼设备。RH 精炼是德国鲁尔钢铁公司和海拉斯公司两家公司在 1957 年共同发明的，故简称 RH 法。

After nearly 40 years of development, RH refining function has been expanding, and it has changed from a simple degassing equipment to a multi-functional processing equipment. The characteristics of its rapid treatment make it match with the fast rhythm of converter. The vacuum circulation treatment is especially suitable for the rapid treatment of large quantities of molten steel, and it can match with the arc furnace or converter with large capacity satisfactorily. Because of the advantages of the vacuum circulation treatment method, it has become one of the fastest developing refining equipment. It can be said that the development of RH far exceeds the expectation of metallurgical workers at the beginning of the invention. It has become a multi-functional refining equipment with degassing, deoxidization, decarbonization, desulfurization, temperature rise, composition control, ultra-low carbon steel refining, stainless steel smelting, etc. And RH has the advantages of simple operation and large amount of steel water treatment. At present, the RH treatment rate of converter molten steel in Japan's five major steel companies (Nippon Steel, Sumitomo, steel pipe, Kobe, and Kawasaki) is more than 70%. The RH process is also developing rapidly in China. The first one was put into operation in Daye Steel Plant in 1968, two RH units were put into operation in Baosteel in 1985 and 1999 respectively, and four RH units were put into operation in WISCO from 1979 to 1990 respectively. The maximum capacity is a 300t RH unit put into operation in December 1985. In recent years, with the increasing proportion of low carbon steel in the market, Rh has been widely used. At present, medium and large steel mills are generally equipped with RH furnaces, and there will be more RH vacuum refining methods in the future.

经过近 40 年的发展，RH 精炼功能不断扩大，已经从一个单纯的脱气设备变成一个多功能的处理设备。其快速处理的特点使它可以与转炉的快节奏配合，真空循环处理特别适用于大批量钢水的快速处理，与大容量的电弧炉或转炉都可以配合。由于真空循环处理方法的优点，使它成为当今发展最快的炉外精炼设备之一。可以说 RH 的发展远远超出了该发明之初冶金工作者对它的期望，其已经成为具有脱气、脱氧、脱碳、脱硫、升温、成分控制、超低碳钢精炼、不锈钢冶炼等多功能精炼设备。而 RH 又具有操作简单、处理钢水量大的优点。现在日本五大钢铁公司（新日铁、住友、钢管、神户、川崎）转炉钢水的 RH 处理比率在 70% 以上。RH 法在我国发展也很快，大冶钢厂 1968 年投产第一台，宝钢分别于 1985 年和 1999 年投产了两台 RH 装置，武钢分别于 1979~1990 年投产了四台 RH 装置，宝钢在 1985 年 12 月投产的一台 300 吨 RH 装置，是最大容量的装置。全国相继建成了 15 台 RH 多功能真空精炼炉，近几年随着低碳钢在市场上所占比例越来越高，RH 的用途越来越广。目前，一般中、大型钢厂都配置有 RH 炉，未来将会有更多的 RH 真空精炼法。

2.11.1　RH Refining Principle and Characteristics

2.11.1　RH 精炼原理及特点

RH Refining is a vacuum circulation degassing method. Its basic process principle is to continuously lift the molten steel to the vacuum chamber for degassing, decarburization and other reactions, and then return it to the ladle. Therefore, RH treatment does not require a specific ladle clearance height, and the reaction speed is not limited by the ladle clearance height.

RH 精炼是真空循环脱气法，其基本工艺原理是利用气泡将钢水不断地提升到真空室内进行脱气、脱碳等反应，然后回流到钢包中。因此，RH 处理不要求特定的钢包净空高度，反应速度也不受钢包净空高度的限制。

The principle of liquid steel vacuum circulation is similar to the function of 'Bubble Pump' (shown in Figure 2-35). When vacuum degassing treatment is carried out, insert two dip tubes at the bottom of the vacuum chamber into the liquid steel at a depth of 100~150mm, and start the vacuum pump to pump the vacuum chamber into a vacuum. Then the vacuum chamber and the shape become a differential pressure, and the liquid steel rises from the two dip tubes to a height with equal differential pressure (high circulation degree). At this time, the liquid steel does not circulate. In order to circulate the liquid steel, argon is blown into the lower part of the riser at the same time. After the argon enters the riser, a large number of bubble cores are generated in the riser due to the thermal expansion. As the volume of the bubble increases hundreds of times when the thermal expansion and pressure decrease, the specific gravity of the liquid steel decreases. And because of the hydrogen and pressure in the argon bubble, the pressure of nitrogen is 0, so the dissolved gas in the molten steel diffuses into the argon bubble. The expanded gas drives the molten steel to rise at a speed of about 5m/s, and then it sprays into the vacuum chamber in the form of a fountain. After the bubble enters the vacuum chamber, it breaks, and the molten steel is broken into small droplets, which greatly increases the degassing area (20~30 times). The gas is separated from the molten steel and pumped away by the vacuum pump, which greatly accelerates the degassing process. The degassed liquid steel is collected at the bottom of the vacuum chamber. Due to the weight difference, it returns to the ladle at the speed of 1~2m/s through the downcomer. The liquid steel without degassing enters the vacuum chamber from the riser and degas again and again, forming a continuous circulation process. After repeated $n(2$~$4)$ cycles, the degassing process is finished.

钢液真空循环原理类似于"气泡泵"的作用,如图 2-35 所示。当进行真空脱气处理时,将真空室下部的两根浸渍管插入钢液内 100~150mm 深度后,启动真空泵将真空室抽成真空,于是真空室内、外形成压差,钢液便从两根浸渍管中上升到压差相等的高度(循环高度)。此时钢液并不循环,为了使钢液循环,从上升管下部约三分之一处吹入氩气,氩气进入上升管内的钢液以后由于受热膨胀,在上升管内瞬间产生大量气泡核,由于该气泡受热膨胀和压力降低时体积成百倍增大,钢液比重变小;由于氩气泡内的氢气和氮气的压力为 0,所以钢液内溶解的气体向氩气泡内扩散,膨胀的气体驱动钢液以约 5m/s 的速度上升,呈喷泉状喷入真空室内。气泡进入真空室后破裂,钢液被碎裂成小液滴,使脱气面积大大增加(20~30 倍),气体自钢液内析出后被真空泵抽走,大大加速了脱气进程。而脱气后的钢液汇集到真空室底部,由于重量的差异,经下降管以 1~2m/s 的速度返回到钢包内。未经脱气的钢液又不断从上升管进入真空室脱气,周而复始,从而形成连续循环过程。如此反复循环 n (2~4) 次后达到脱气目的,脱气过程结束。

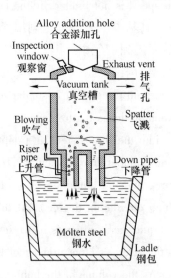

Figure 2-35　RH refining
图 2-35　RH 精炼

The main features of RH are as follows:
RH 的主要特点如下:

(1) The treatment cycle is short and the production efficiency is high. It is often used together with converter.

(1) 处理周期短,生产效率高,常与转炉配套使用。

(2) With high reaction efficiency, the molten steel can react directly in the vacuum chamber, and can produce ultra pure steel with $w[H] \leq 0.5 \times 10^{-4}\%$, $w[N] \leq 25 \times 10^{-4}\%$, and $w[C] \leq 10 \times 10^{-4}\%$.

（2）反应效率高，钢水直接在真空室内进行反应，可生产 $w[H] \leq 0.5 \times 10^{-4}\%$，$w[N] \leq 25 \times 10^{-4}\%$，$w[C] \leq 10 \times 10^{-4}\%$的超纯净钢。

(3) It can be used for oxygen blowing decarbonization and secondary combustion for heat compensation to reduce treatment temperature drop.

（3）可进行吹氧脱碳和二次燃烧进行热补偿，减少处理温降。

(4) It can be used to produce ultra-low sulfur steel with $w[S] \leq 5 \times 10^{-6}$.

（4）可进行喷粉脱硫，生产 $w[S] \leq 5 \times 10^{-6}$的超低硫钢。

Compared with other refining technologies, the advantages of RH refining include good degassing effect. In the process of RH treatment, the amount of liquid steel entering the vacuum chamber is relatively small, and a large number of bubbles are generated in the riser due to the driving gas blown in. Part of the liquid steel entering the vacuum chamber is in a small droplet and boiling state, which increases the degassing surface area of the liquid steel, and which is conducive to degassing. It is suitable for the treatment of a large number of liquid steel with large production capacity, short treatment cycle and treatment process, and the temperature drop is small. In the process of treatment, the surface of molten steel in the ladle is covered by slag, which has good heat preservation effect. Generally, the temperature drop after treatment is only 30~50℃; The scope of application is relatively large, and the same equipment can be used to treat molten steel with different capacities, which is suitable for various levels of steelmaking processes such as converter, electric arc furnace, induction furnace, etc.

与其他精炼技术相比，RH精炼的优点包括脱气效果好。RH处理过程中，进入真空室的钢液量相对较少，而且由于吹入的驱动气体在上升管内生成大量气泡，进入真空室的部分钢液呈细小的液滴，且处于沸腾状态，增大了钢液脱气表面积，有利于脱气的进行，适用于大量钢液的处理，生产能力大；处理周期短，处理过程温降小。处理过程中钢包内的钢液表面有炉渣覆盖，保温效果好，一般处理后温降仅为30~50℃；适用范围较大，用同一设备能处理不同容量的钢液，适用于转炉、电弧炉、感应炉等各种级别的炼钢工艺。

2.11.2 RH Refining Equipment

2.11.2 RH精炼设备

RH refining equipment is composed of degassing main equipment, water treatment equipment, electrical equipment and instrument equipment. The equipment is shown in Figure 2-36.

RH精炼设备由脱气主体设备、水处理设备、电气设备、仪表设备所组成。其设备示意图如图2-36所示。

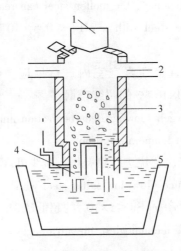

Figure 2-36 RH refining equipment
图 2-36 RH 精炼设备
1—Alloy adding hole; 2—Exhaust hole; 3—Vacuum chamber; 4—Riser; 5—Downcomer
1—合金添加孔; 2—排气孔; 3—真空室; 4—上升管; 5—下降管

2.11.3 RH Refining Process RH

2.11.3 RH 精炼工艺过程

In the long process of modern iron and steel production, the best process of modern converter steelmaking is: Blast furnace → Pretreatment → Converter of top and bottom combined blowing (BOF) → RH vacuum refining (or CAS-OB refining) → Continuous casting → Rolling (or continuous) → Casting billet hot delivery → Direct rolling.

在现代化钢铁生产长流程中，现代转炉炼钢生产的最佳工艺流程为：高炉→铁水炉外预处理→转炉顶底复合吹炼（BOF）→RH 真空精炼（或 CAS-OB 精炼）→连铸连轧（或连铸）→铸坯热送→直接轧制。

The RH Refining process usually involves temperature measurement and sampling of the ladle after tapping. It lower the vacuum chamber with the insertion depth of 150~200mm and start the vacuum pump and inputting the driving gas into one insertion tube. When the pressure of the vacuum chamber drops to 26~10kPa, the circulation is intensified; The rising speed of the molten steel is 5m/s and the falling speed is 1~2m/s. The gas and inclusions are brought out by bubbles in the molten steel. It is divided into two parts: preparation before refining and vacuum treatment.

精炼工艺通常在出钢后，钢包测温取样；下降真空室，插入深度为 150~200mm；起动真空泵，一根插入管输入驱动气体；当真空室的压力降到 26~10kPa 后，循环加剧；钢水上升速度为 5m/s、下降速度为 1~2m/s；气泡在钢液中将气体及夹杂带出。该工艺具体分为精炼前的准备和真空处理两个环节。

Preparation before refining includes supply of vacuum chamber baking, electricity, compressed air, water vapor and cooling water, preparation of driving gas and reaction gas, and preparation of alloy hopper. After confirming that the preparatory work is in good condition, it can be put into operation, the baking is stopped, the slag blocking cap is put on the lifting pipe mouth, and the alloy hopper is installed on the vacuum chamber. At the same time, the primary smelting furnace is out of steel, the steel ladle is transported to the lifting trolley, and the temperature is measured and sampled in the ladle.

精炼前的准备包括：真空室烘烤、电、压缩空气、水蒸气、冷却水的供应，驱动气体和反应气体的准备，以及合金料斗的准备。确认准备工作完好后即可投入运行，停止烘烤，在升降管口套上挡渣帽，在真空室上装合金料斗。同时，初炼炉出钢，钢包运至升降台车，在钢包中测温取样。

Then lower the vacuum chamber, insert the lifting tube into the liquid steel, the depth is not less than 150~200mm. The vacuum pump is started with the decrease of the pressure of the vacuum chamber, and the liquid steel rises along the lifting tube. When the driving gas is blown into the lifting tube, the pressure of the vacuum chamber drops to 26~13kPa, the liquid steel obviously circulates through the vacuum chamber, splashes in the vacuum chamber, and the surface area increases significantly. Thus the degassing process is accelerated. The degassed liquid steel is collected at the bottom of the vacuum chamber. Under the action of gravity, the liquid steel returns to the ladle at a speed of 1~2m/s. The degassed liquid steel is impacted to mix and mix with each other. After several cycles, the gas in the liquid steel can be reduced to a relatively low level.

然后降下真空室，将升降管插入钢液内，深度不小于150~200mm。启动真空泵，随着真空室压力的降低，钢液沿着升降管上升。当向上升管吹入驱动气体，真空室压力降到26~13kPa时，钢液明显地经过真空室循环，并在真空室内喷溅，表面积显著增大，从而加速脱气过程，脱气后的钢液汇集到真空室底部。在重力作用下，以1~2m/s的速度返回钢包，冲击未脱气的钢液，使其相互搅拌和混合，经过若干次循环后，可将钢液内的气体降到相当低的水平。

Temperature measurement and sampling shall be conducted every 10min at the initial stage of the cycle, and every 5min at the end of the process. According to the results of sampling analysis, if alloy materials need to be added, the hopper with automatic control can be operated. Under the condition of not damaging the vacuum degree, the alloy materials shall be added into the vacuum chamber at a constant feeding speed, and recycled for several minutes after the material is added, so as to ensure the temperature uniformity of the components and the treatment. After completion, when the vacuum pump is closed to lift the vacuum chamber, temperature measurement and sampling shall be conducted again. After the composition is qualified, the ladle shall be removed for pouring.

循环初期每隔10min测温、取样一次，接近处理终点时每隔5min测温、取样一次。根据取样分析结果，如需补加合金料，可操作自动控制的料斗，在不破坏真空度的条件下，以恒定的加料速度将合金料加入真空室内。料加完后再循环几分钟，以保证成分温度均匀。处理完毕后，在关闭真空泵提升真空室的同时，再测温取样，成分合格后将钢包移出进行浇注。

2.11.4 RH Refining Vacuum Treatment

2.11.4 RH 精炼真空处理

The vacuum treatment of RH process is divided into present treatment and light treatment. RH light treatment process is to use the mixing and decarburization function of RH to treat the non deoxidized molten steel for a short time under the condition of low vacuum, and adjust the temperature and composition of the molten steel to meet the process requirements of continuous casting.

RH 法真空处理分为本处理和轻处理。RH 轻处理工艺就是利用 RH 的搅拌、脱碳功能，在低真空条件下，对未脱氧钢水进行短时间处理，同时将钢水温度、成分调整到适于连铸的工艺要求。

2.11.4.1 Present Treatment

2.11.4.1 本处理

Present treatment refers to the vacuum degassing treatment under high vacuum (pressure less than 270Pa) for the purpose of removing hydrogen and oxygen (deoxidizing products) in molten steel. Because dehydrogenation is the most traditional and mature treatment process of RH, it is called the fundamental treatment of RH (present treatment). Generally, the dehydrogenation rate is about 50%, and the maximum is 75%. It is required that [H] of RH treatment is less than 7ppm.

本处理是指在高真空下（压力小于 270Pa），以去除钢水中的氢、氧（脱氧生成物）为目的的真空脱气处理。因脱氢处理是 RH 最传统最成熟的处理工艺，故称 RH 的根本处理（本处理）。一般脱氢率在 50% 左右，最大可达 75%，要求 RH 处理 [H] 含量小于 7ppm。

The characteristics of present treatment process are as follows:

本处理工艺特点包括：

(1) The main purpose of this treatment is dehydrogenation.

(1) 本处理的主要目的是脱氢。

(2) The molten steel must be fully killed steel, $f[O]<5ppm$.

(2) 钢水必须是完全脱氧镇静钢，且 $f[O] \leqslant 5ppm$。

(3) The whole pump is put into operation quickly and treated under high vacuum for a certain time.

(3) 迅速全泵投入，在高真空度下处理一定时间。

(4) Increase the circulation and accelerate the floating speed of inclusions.

(4) 加大循环量，加快夹杂的上浮速度。

(5) The refractory for ladle and vacuum tank must reach a certain baking temperature and use times and be used continuously.

(5) 钢包、真空槽耐火材料必须达到一定的烘烤温度和使用次数并连续使用。

(6) The end point [H] of RH is less than 2ppm, which is generally required to be less than 1.5ppm to avoid white spot defects.

(6) RH 终点 [H] 不小于 2ppm，现一般要求小于 1.5ppm，避免白点缺陷。

(7) In order to ensure the effect of dehydrogenation, the alloy should basically reach the lower limit of steel grade before RH treatment.

(7) 为保证脱氢效果，在 RH 处理前合金应基本达到钢种下限。

(8) Cr, Mn, Ti and Nb in steel increase the solubility of hydrogen, while C, Si and Al decrease the solubility.

(8) 钢中 Cr、Mn、Ti、Nb 增加氢的溶解度，C、Si、Al 降低溶解度。

(9) The main factors affecting dehydrogenation are original hydrogen content, raw material type, state, quantity, climate, refractory, deoxidation degree of steel, vacuum degree and treatment time.

(9) 主要影响脱氢的因素有原始氢含量、原材料品种、状态、数量、气候、耐火材料、钢脱氧程度、真空度和处理时间。

(10) Present treatment is applicable to low alloy structural steel, cold rolled plate steel, petroleum pipeline steel, high pressure vessel steel, low carbon steel, SPAH, X42, X70, 19Mn6, Q195 and 08Al.

(10) 本处理适用钢种包括低合金结构钢、冷轧板钢、石油管线钢、高压容器钢、低碳钢、SPAH、X42、X70、19Mn6、Q195 和 08Al。

2.11.4.2 Light Treatment

2.11.4.2 轻处理

In 1977, a new RH treatment process, called RH light treatment process, was developed by Nippon Steel Corporation. The basic process is to deoxidize in unkilled steel or semikilled steel smelted by converter at 40~20kPa (300~150 Torr) for about 10min, then add deoxidizer at 6666~1333Pa (50~10 Torr) for about 2min, and fine tune it to make the composition and temperature of liquid steel reach the most suitable conditions for continuous casting. Compared with general RH treatment, its treatment time is shorter, energy consumption and temperature drop are lower, and the yield of ferroalloy is improved. For example, [O] is about 400ppm lower than that of common method (no treatment) when adding aluminum for deoxidization. So the yield of aluminum is improved significantly.

1977 年，日本新日铁大分厂研究出了一种新的 RH 处理工艺，叫 RH 轻处理工艺。其基本过程为：将转炉冶炼的未脱氧钢或半脱氧钢，先在 40~20kPa（300~150 托）的真空度下碳脱氧（10min 左右），然后在 6666~1333Pa（50~10 托）下加脱氧剂脱氧（约 2min），并进行微调，使钢液成分和温度达到最合适连铸的条件。与一般 RH 处理相比，它的处理时间短，能耗和温降都较低，铁合金收得率提高。如加铝脱氧时，[O] 比普通方法（不处理）低，约 400ppm。因而铝的收得率显著提高。

The light treatment process is particularly suitable for the production of quasi boiling steel. Under the low vacuum degree of RH vacuum chamber, the non deoxidized liquid steel reacts with [C] and [O] in the liquid steel to generate co bubbles, so as to reduce the oxygen content. The content of Al is limited in a certain range. In order to control the content of Al accurately, the oxygen content in molten steel is usually determined by oxygen meter, and then the aluminum content is added according to the situation.

轻处理工艺特别适于准沸腾钢生产。不脱氧钢液在 RH 真空室内的低真空度下，使钢液中的 [C] 和 [O] 产生反应生成 CO 气泡，以减少氧的含量。Al 含量限制在一定的范围内，为了准确地控制 Al 含量，通常利用定氧仪来测定钢液中的含氧量，然后再视情况决定添加 Al 含量。

The key points of light treatment operation are as follows:

轻处理操作要点包括：

(1) The vacuum degree shall be controlled at 6~15kPa. The vacuum degree shall be adjusted properly according to the splashing of molten steel in the vacuum tank.

(1) 真空度控制在 6~15kPa，处理时根据真空槽内钢水的飞溅情况适当调整真空度。

(2) Pure degassing time (except for oxygen blowing, between the end of the last batch of alloy addition and the end of treatment, pure steel circulation, uniform composition and degassing time) is more than 3min.

(2) 纯脱气时间（除吹氧以外，最后一批合金加完到处理结束之间，单纯进行的钢水循环，均匀成分和脱气时间）不少于 3min。

(3) The amount of decarburization in the light treatment of low carbon Al killed steel is about 0.04%. When the converter tapping [C] is too high, the decarburization can be carried out by top gun blowing.

(3) 低碳铝镇静钢轻处理中的脱碳量约 0.04%。当转炉出钢 [C] 过高时，可以用顶枪吹氧脱碳。

Task 2.12　AOD Refining
任务 2.12　AOD 精炼

Since the 1950s, with the increasing use of low price and high carbon ferrochromium, in order to produce a series of stainless steel, smelting temperature must be increased to reduce the burning loss of chromium under normal pressure. As a result, the service life of furnace lining is greatly reduced. In the first half of the 1960s, the partial pressure of CO was reduced by means of using O_2 to decarbonize or diluting gas under vacuum to achieve decarburization and chromium retention. The AOD furnace (Argon Oxygen Decarburization) was born. It is a patent of krewski of United Carbide Company of America. In 1968, the first AOD furnace (15t) was built in the world. At present, AOD method is mainly used to produce stainless steel in the world, accounting for 75% of the total output. In addition, VOD method accounts for 15%, and the remaining 10% are produced by CLU method, ASEA-SKF method, RH-OB method and LFV method. The world stainless steel manufacturers quickly accept AOD refining. Taiyuan Iron and Steel Co., Ltd. in China built the first AOD furnace in 1989.

自20世纪50年代以来，随着低价格高碳铬铁使用的增加，为减少铬的烧损，在常压下生产一系列不锈钢种，就必须提高冶炼温度，其结果是极大降低了炉衬的使用寿命。20世纪60年代上半期采用通过真空下用O_2脱C，或用稀释气体的方法来降低CO分压达到去碳保铬，AOD炉（即氩氧脱碳法）就此诞生，它是美国联合碳化物公司的克里夫斯基的一项专利。1968年，乔斯琳钢公司在世界上建成第一台AOD炉（15t）。目前，世界上主要采用AOD法生产不锈钢，占总产量的75%。另外，VOD法占15%，其余10%应用CLU法、ASEA-SKF法、RH-OB法、LFV法所生产。世界不锈钢厂家能够迅速接受AOD精炼，中国太原钢铁在1989年建成国内第一台AOD炉。

2.12.1　Refining Equipment of AOD

2.12.1　AOD精炼设备

The main structure of AOD furnace is composed of furnace body, gas supply device, dust removal device and feeding device. The shape of the oxygen converter is similar to that of a straight cylinder. It is composed of a steel plate enclosed furnace shell and a lining lined with refractories. The service life of the refractory lining is short. The movable shell furnace body can rotate forward or backward 180°, and the tilting speed is generally divided into two types: low speed is used in smelting operation, and high speed is used in other operations. The inner shape of the furnace is a round table body, which is composed of a cylinder in the middle of the furnace body and the inverted round table body in the lower part. The ratio of inner diameter to height to the general proportion is 1:2:3, as shown in Figure 2-37.

炉的炉体的主要由炉体、供气装置、除尘装置和加料装置四部分组成。其炉体形状近似直筒形氧气转炉，是由钢板围制的炉壳和内砌耐火材料的炉衬组成。耐火炉衬的寿命较短，活动炉壳炉体可以向前或向后旋转180°。倾动速度通常分为两种：冶炼操作时使用低速，其他作业时选用高速。炉膛的内形为圆台体，炉体中部的圆柱体和下部的倒置由圆台体所组成。熔池深度、内径和高度的一般比例为1:2:3。AOD炉结构示意图如图2-37所示。

In addition to the mechanical tilting device of the furnace body and the metal structure supporting the device, the AOD refining furnace also has a feeding system composed of material bin, weighing, conveying and putting into the furnace. The system can automatically add reductant, desulfurizer, ferroalloy or coolant etc., according to the technological requirements during the blowing process. The temperature measurement and sampling system can ensure the temperature measurement and sampling during the blowing process, and record the measurement results in the operation room. In addition, there are drying and preheating devices for the furnace body.

AOD精炼炉除炉体的机械倾动装置和支撑该装置的金属结构外，还有料仓、称量、

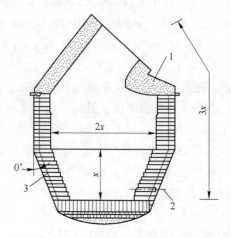

Figure 2-37　AOD furnace structure
图 2-37　AOD 炉结构示意图
1—Furnace cap; 2—Tuyere; 3—Furnace bottom
1—炉帽; 2—风口; 3—炉底

输送和投入炉内等部件组成的加料系统。该系统可以在吹炼过程中，按工艺要求自动加入还原剂、脱硫剂、铁合金或冷却剂等。测温取样系统可以保证在吹炼过程中测温取样，并在操作室中记录测量结果。此外还有炉体的干燥和预热装置等。

The air inlet for blowing in the mixed gas for refining is installed on the side wall opposite to the side of the tapping port and close to the furnace bottom. The type of AOD furnace tuyere is unique. It is a consumable tuyere cooled by gas. The tuyere adopts a double-layer casing structure, the outer tube is only filled with argon to cool the tuyere, and the inner tube is filled with argon oxygen mixture, as shown in Figure 2-38.

吹入精炼用混合气体的风口，安装在出钢口侧对面、靠近炉底的侧壁上。AOD 炉风口的型式是特有的，它是用气体冷却的消耗式风口。风口采用双层套管结构，其外管只通氩气来冷却风口，内管通氩氧混合气体。AOD 炉风口示意图如图 2-38 所示。

Figure 2-38　Schematic diagram of AOD furnace tuyere
图 2-38　AOD 炉风口示意图

2.12.2　Principle of AOD

2.12.2　AOD 法工作原理

The AOD method blows in argon oxygen mixture from the side of furnace bottom to the molten pool. Blow in inert gas (Ar, N_2) as dilution gas, reduce the partial pressure p_{CO} of reaction product CO, achieve the purpose of decarburization and chromium retention. In the blowing process, oxygen of 1mol reacts with the carbon in the steel to form Co of 2mol. But argon of 1mol does not change after passing through the molten pool, so that the partial pressure of CO in the upper part of the molten pool is reduced, which is greatly conducive to the decarburization and chromium retention in the smelting of stainless steel.

AOD 法以氩氧混合气体的形式从炉底侧面向熔池中吹入。吹入惰性气体（Ar、N_2）作稀释气体，从而降低反应产物 CO 的分压 P_{co}，以达到脱碳保铬的目的。在吹炼过程中，1mol 氧气与钢中的碳反应生成 2mol CO，但 1mol 氩气通过熔池后没有变化，从而使熔池上部 CO 的分压力降低，从而有利于冶炼不锈钢时的脱碳保铬。

2.12.3 AOD Process

2.12.3 AOD 法工艺过程

After the primary smelting furnace is discharged, the liquid steel is weighed and slag is removed during the process of transportation to AOD furnace, and then mixed into AOD furnace. In order to prevent liquid steel from pouring into the tuyere, argon or nitrogen will be blown into the tuyere when mixing liquid steel. According to the content of carbon, silicon, manganese and the quality of molten steel, the total amount of oxygen blown in is calculated. In order to reduce the oxidation of chromium and prevent the high temperature of molten steel, the mixing ratio of argon and oxygen should be changed during blowing. In practical operation, argon oxygen ratio can be changed in three or four stages. The mixing ratio used in each stage shall be determined according to the carbon content and temperature at the end of the stage. In order to protect the refractories of the furnace body, the temperature of the molten steel should not exceed 1750℃, so the clean and dry return steel of the same steel grade can be added in the oxidation period to cool the molten steel. In addition, the use of nickel oxide in raw materials is beneficial to shorten the oxidation time, and reduce the consumption of blowing gas and the cost of raw materials.

初炼炉出钢后，在运往 AOD 炉的过程中，对钢液称量和除渣，然后兑入 AOD 炉中。为防止钢液灌入风口，在兑钢液时由风口吹入氩气或氮气。根据初炼炉出钢的碳、硅、锰含量和钢液的质量，计算出吹入氧的总量。为了尽量减少铬的氧化和防止钢液温度过高，在吹炼时要改变氩氧的混合比。在实际操作中，分三或四个阶段变化氩氧比例。每个阶段所用的混合比按该阶段终了时的含碳量和温度来确定。为了保护炉体耐火材料，钢液温度不宜超过1750℃，因而在氧化期可加入清洁干燥的同钢种的返回钢来冷却钢液。此外，在原料中使用氧化镍，有利于缩短氧化时间，减少吹炼用气体的消耗和降低原材料费用。

In the mixed gas used for AOD blowing, the volume ratio of oxygen to argon is called the ratio of oxygen and argon. In 1700℃, when $w[Cr] = 18\%$, pure oxygen can be blown into the stage when $w[C] > 0.20\%$. When $w[C]$ is less than 0.20%, the proportion of argon will increase gradually.

AOD 吹炼所用的混合气体中，氧气与氩气的体积比称为氧氩比。在 1700℃，$w[Cr] =$

18%条件下，当$w[C]>0.20\%$时，可以吹入纯氧气；当$w[C]<0.20\%$时，再逐渐增加氩气的比例。

The first stage is that supply gas according to the proportion of $w(O_2):w(Ar)=3:1$, and reduce carbon to about 0.20%. At this time, the temperature is about 1680℃.

在第一阶段，按$w(O_2):w(Ar)=3:1$比例供气。当碳（质量分数）降低到0.20%左右时，这时的温度约为1680℃。

The second stage is that according to the ratio of $w(O_2):w(Ar)=2:1$, the gas supply is carried out, and the carbon is reduced to about 0.10%. At this time, the temperature can be as high as 1740℃.

在第二阶段，按$w(O_2):w(Ar)=2:1$比例供气。当碳（质量分数）降低到0.10%左右时，这时的温度高达1740℃。

The third stage is that supply gas according to the ratio of $w(O_2):w(Ar)=1:2$, and reduce the carbon to the required limit.

在第三阶段，按$w(O_2):w(Ar)=1:2$比例供气，将碳降到所需要的极限。

The main purposes of the first and second stages are decarburization, desilication, desulfurization and adjustment of the composition of the molten steel. In the third stage, CO partial pressure drops below 10kPa, decarburization continues and slag is adjusted to reduction slag to recover chromium from slag.

第一阶段和第二阶段的主要目的是脱碳、脱硅、脱硫和调整钢液成分。第三阶段CO分压降至10kPa以下，继续脱碳，将炉渣调整为还原渣，回收渣中的铬。

When the oxygen to argon ratio has been determined, the flow rate of the mixed gas only depends on the flow rate of oxygen. In AOD blowing, the oxygen demand is determined according to the composition of molten steel. Considering the most reasonable decarburization rate and time, the oxygen flow rate can be determined.

在氧氩比确定的情况下，混合气体的流量只取决于氧气的流量。在AOD吹炼中，氧气的需用量是按入炉钢液的成分确定的。考虑到最合理的脱碳速率和吹氧脱碳的时间，即可确定氧气的流量。

Exercises

练 习 题

(1) What is LF method? And what are the main advantages of LF Process?

(1) 什么是LF法，LF工艺的主要优点有哪些？

(2) What are the main equipment of LF method?

(2) LF法的主要设备包括哪些？

(3) What is the basic function and composition of LF refining slag?

(3) LF精炼渣的基本功能是什么，其组成包括哪些？

(4) Why is argon blowing in the whole process of VD? And what is the requirement of ladle clearance for VD refining?

(4) VD 法处理过程为什么要全程吹氩，VD 法精炼对钢包净空有什么要求？

(5) Discussed the working principle of RH method and DH method.

(5) 试述 RH 法与 DH 法的工作原理.

(6) What is the basic equipment of 3-2 RH process? And what is its metallurgical function and smelting effect?

(6) 3-2 RH 法的基本设备包括哪些部分，其冶金功能与冶炼效果如何？

Project 3　Basic Knowledge of Refractories for Secondary Refining

项目3　炉外精炼用耐火材料基础知识

All kinds of means used in the process of refining outside the furnace are extremely harsh for the service conditions of equipment, especially the containers used in refining. Due to the unique characteristics of refining, the refractory of refining vessel (mostly ladle) is required to be very high, and its refractory has a very important impact on the safety production of smelting. Therefore, refractories can be used as the basic guarantee of refining technology outside the furnace. Metallurgical workers usually compare refractories to the mother of iron and steel production.

炉外精炼过程中所采用的各种手段，对于设备的使用条件是极其苛刻的，特别是精炼所使用的容器。由于精炼的独特决定了对精炼容器（绝大多数是钢包）的耐火材料要求很高，其耐火材料对冶炼的安全生产有着非常重要的影响。因此耐火材料可以作为炉外精炼技术的基本保障，冶金工作者通常把耐火材料比作钢铁生产之母。

Task 3.1　Requirements of Refractories for Secondary Refining

任务3.1　炉外精炼对耐火材料的要求

During refining, the working conditions of the ladle are as follows:

精炼时钢包的工作条件如下：

(1) high temperature: When oxygen blowing and arc heating, the temperature is as high as 1600~1800℃, and the local overheating of the ladle lining is near the arc.

(1) 高温：当吹氧和电弧加热时，温度高达1600~1800℃，电弧附近包衬局部过热。

(2) In the process of circulation degassing, argon blowing and electromagnetic stirring, the molten steel and slag scour and erode the furnace lining violently.

(2) 在循环脱气、吹氩和电磁搅拌过程中，钢液和熔渣对炉衬剧烈冲刷和侵蚀。

(3) Slagging: The impurities are removed with composition adjustment, so as to prolong the retention time of the molten steel in the ladle.

(3) 造渣：去除杂质和成分调整，使钢液在包内的停留时间延长。

(4) The thermal stress at high temperature and the temperature change caused by intermittent operation. So the working conditions of the refining furnace outside the furnace are worse than those of the conventional furnace.

(4) 高温时的热应力和间歇式的工作引起温度剧变，因此炉外精炼炉较常规炉的工作条件恶劣。

Therefore, refractories must be selected according to the performance, treatment method, smelting steel and operation level of the refining equipment outside the furnace. In addition, due to the different damage mechanism of the lining in each part, the corresponding materials should also be selected to build the furnace comprehensively, so as to improve the life of the furnace and reduce the cost. The requirements of refractories for secondary refining are as follows:

所以，必须根据炉外精炼设备的性能、处理方法、冶炼钢种和操作水平来选择耐火材料。同时，由于各部位内衬损毁机理不同，也应分别选择相应的材质，进行综合砌炉，以达到提高窑炉寿命、降低成本的目的。炉外精炼工对耐火材料的要求包括：

（1）They have high fire resistance and good stability, and can resist the high temperature and vacuum effect under the condition of refining outside the furnace.

（1）耐火度高，稳定性好，能抵抗炉外精炼条件下的高温与真空作用。

（2）The porosity is low, the bulk density is large, and the structure is compact, so as to reduce the infiltration of slag.

（2）气孔率低，体积密度大，组织结构致密，以减少炉渣的浸透。

（3）They have high strength and wear-resistant, and can resistant to steel slag erosion.

（3）强度大，耐磨损，能抵抗钢渣冲刷磨损。

（4）They have good corrosion resistance and can resist the erosion of acid alkali slag.

（4）耐侵蚀性好，能抵抗酸—碱性炉渣的侵蚀作用。

（5）They have good thermal stability, no thermal shock cracking and peeling.

（5）热稳定性好，不发生热震崩裂剥落。

（6）They do not pollute the molten steel and beneficial to the purification of the molten steel.

（6）不污染钢液，有利于钢液的净化作用。

（7）The pollution to the environment is small.

（7）对环境的污染小。

Task 3.2　Varieties and Types of Refractory Materials for External Refining

任务 3.2　炉外精炼用耐材的品种与类型

Refining outside the furnace is mainly carried out in the ladle and AOD furnace (AOD furnace is similar to converter). In the past, refining ladle lining mainly used high alumina, quartz, magnesia chrome and dolomitic materials. Now, it has developed to use MgO-C, MgO-CaO-C and spinel refractories. Generally, dolomite, magnesia dolomite and magnesia carbon refractories are used in AOD furnace. At present, high alumina, dolomite and carbon products are mainly used in the ladle for refining outside the furnace. And magnesium chromium products have been gradually replaced by carbon products. The varieties and types of ladle refractory used in China are as follows:

炉外精炼主要是在钢包内和 AOD 炉内进行（AOD 炉类似于转炉）。以往精炼钢包炉衬主要使用高铝质、锆英石质、镁铬质和白云质材料，如今已发展到使用 MgO-C、MgO-CaO-C 和尖晶石质耐火材料。AOD 炉一般使用白云石质、镁白云石质和镁碳质耐火材料。目前，炉外精炼所用钢包主要使用高铝质、白云石质和碳质制品，镁铬质制品已逐渐被含碳制品所取代。我国炉外精炼使用的钢包耐材的品种和类型包括：

(1) Ladle lining of alumina magnesia brick is the main of refractory material;

(1) 以铝镁碳砖为主要耐火材料的铝镁砖钢包内衬；

(2) Dolomite brick ladle lining with dolomite brick as the main refractory;

(2) 以白云石砖为主要耐火材料的白云石砖钢包内衬；

(3) The lining of ladle made of alumina magnesia spinel as the main refractory;

(3) 以铝镁尖晶石浇筑料为主要耐火材料的铝镁尖晶石浇筑钢包内衬；

(4) MgO-CaO-C brick ladle lining;

(4) MgO-CaO-C 砖钢包内衬；

(5) Full MgO-C brick ladle lining;

(5) 全 MgO-C 砖钢包内衬；

(6) Magnesia chrome brick ladle lining.

(6) 镁铬砖钢包内衬。

The data show that the proportion of the cost of refining refractories outside the furnace in the operation cost increases greatly because of the low life of ladle lining. For example, for VAD furnace, refractory cost accounts for 40%. For LF furnace, it accounts for 53%. After many years of research and practice, the selection of refractories for furnace refining and lining manufacturing process has made great progress. For example, the service life of direct combined magnesia chrome brick for RH, dolomite brick or magnesia carbon brick for LF slag line, high-purity dolomite carbon brick and magnesia dolomite brick for VAD and ASEA-SKF furnaces, fired magnesia magnesia brick, dolomite brick, magnesia dolomite brick and magnesia chrome brick for stainless steel refining vessels have been multiplied. However, compared with the lining life of converter and electric arc furnace, the service life of the lining of refining equipment outside the furnace is still very low, especially in some parts, such as the slag line of LF furnace, the lining of VOD and AOD furnace, the area around the vent of permeable brick and AOD, and the riser of RH degassing device. Because of the damage of the refractory lining in these parts, the whole equipment will stop running or be forced to replace the whole lining, which not only affects the production but also increases the cost. The average service life of direct combined magnesia chrome brick used in AOD furnace in Japan is 50 times, and the service life of magnesia dolomite brick is 70 times. In China, the vulnerable parts are combined with magnesia chrome bricks, with a maximum service life of 34 times. Oil impregnated magnesia dolomite brick is used, with a maximum service life of 46 times. In the slag line of TN process, MgO dolomite brick is used in Japan. After 2~5 times of minor repair, its service life reaches 90~120 times. Therefore, it is necessary to further research and develop the material, manufacturing process, use specification and damage mechanism of refractories used in the refining equipment outside the furnace.

资料表明，因为钢包衬寿命不高，导致炉外精炼耐火材料费用在操作费用中的比例增加。例如对于 VAD 炉，耐火材料费用占 40%；对于 LF 炉，占 53%。经过多年的研究实践表明，用于炉外精炼的耐火材料的选择和内衬制作工艺有较大进展。如在 RH 运用的直接结合的镁铬砖，在 LF 炉渣线运用的白云石砖或镁碳砖，在 VAD 和 ASEA-SKF 炉运用的高纯度白云石碳砖和镁白云石砖，在不锈钢精炼容器运用的烧成铬镁砖、白云石砖、镁白云石砖和镁铬砖，其寿命都成倍地增长。但是相对于转炉和电弧炉炉衬寿命，炉外精炼设备内衬的使用次数仍然很低，特别在个别部位，如 LF 炉的渣线，VOD、AOD 炉的内衬，透气砖和 AOD 风口周围，以及 RH 脱气装置的上升管等，都是使用寿命特别低的部位。由于这些部位耐火内衬的损坏会导致整个设备停止运行，或被迫更换整个内衬，既影响了生产又增加了费用。我国的易损部位使用再结合镁铬砖，最高使用寿命 34 次；采用油浸镁白云石砖，最高使用寿命 46 次。在 TN 法上的渣线部位，日本使用镁白云石砖，经过 2~5 次中小修，其寿命达 90~120 炉次。因此，对炉外精炼设备使用的耐火材料的材质、制作工艺、使用规范，以及损坏的机理等问题，需要继续深入研究与开发。

Exercises

练 习 题

（1）What are the requirements for using refractories in out of furnace refining?
（1）炉外精炼对使用耐火材料的要求有哪些？
（2）What kinds of refractories are used for refining outside the furnace?
（2）炉外精炼用耐火的品种和类型有哪些？

References
参 考 文 献

[1] Li Maowang, Hu Qiufang. *Secondary Refining* [M]. Beijing: Metallurgical Industry Press, 2016.

[2] Zhang Shixian, Zhao Xiaoping, Guan Xin. *The technology of Secondary Refining* [M]. Beijing: Metallurgical Industry Press, 2013.

[3] Zhang Yan. *Learning guidance for refining workers of Secondary Refining* [M]. Beijing: Metallurgical Industry Press, 2015.

[4] Gao Zeping, He Daozhong. *Secondary Refining* [M]. Beijing: Metallurgical Industry Press, 2005.

[5] Yu Haiming. *Secondary Refining in EAF* [M]. Beijing: Metallurgical Industry Press, 2010.

[6] Chen Jianbin. *Secondary Treatment* [M]. Beijing: Metallurgical Industry Press, 2008.

[7] Li Maowang, Hu Qiufang. *Practical Training Guide Book of Secondary Refining* [M]. Beijing: Metallurgical Industry Press, 2016.

[8] Zhou Lanhua, Xia Yuhong. 500 *Questions on Secondary Refining* [M]. Beijing: Chemical Industry Press, 2010.

[9] 鞍钢科学技术馆钢铁情报研究所编. 钢的炉外精炼文集 [M]. 1978.

[10] Shanghai Iron and Steel Research Institute, et al. *Translation of Secondary Refining* [M]. Architectural Research Institute, Ministry of metallurgical industry, 1979.

[11] Wang Chengxun. *Refractories for Secondary Refining* [M] (2nd). Beijing: Metallurgical Industry Press, 2007.

[12] Bo Xing. *The Challenge of Secondary Refining to Mass Production of Multi Variety and High Quality Steel* [M]. Beijing: Metallurgical Industry Press, 2002.

[13] Bo Xing, Wei Gang, Li Hong. *Secondary Refining* [M]. Beijing: Metallurgical Industry Press, 2002.

[14] Qiu Yike, H. M. *Method of Improving Steel Quality in Secondary Refining* [M]. Beijing: Metallurgical Industry Press, 1983.

[15] Li Weili. *Foreign Refining Technology Secondary Refining* 1 [M]. Refining office outside the furnace, Ministry of metallurgy, 1993.

[16] Zhao Wei. *Practical Technical Manual for Secondary Refining and Hot Metal Pretreatment* [M]. Beijing: Metallurgical Industry Press, 2004.

[17] Onur Aydemir. Use of aluminium dross for slag treatment in secondary steelmaking to decrease amount of reducible oxides in ladle furnace [D]. Turkey, Middle East Technical University, 2007.

[18] A. Ghosh. *Secondary Steelmaking——Principles and Applications* [M]. CRC, 2011.